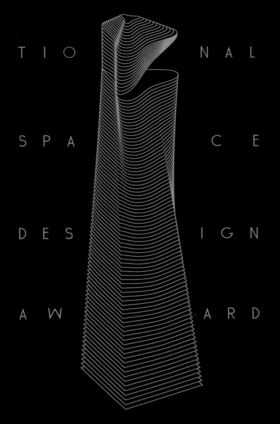

INTERNATIONAL SPACE DESIGN AWARD

Idea-Tops 艾特奖

2015 艾特奖获奖作品年鉴
国际空间设计大奖艾特奖组委会 编著

中国水利水电出版社
www.waterpub.com.cn
·北京·

2015 艾特奖获奖作品

THE AWARD WINNING IDEA-TO

郑时龄
中国科学院院士、
同济大学博士生导师

Zheng Shiling, Academician of Chinese Academy of Sciences and Doctoral Supervisor of Tongji University

我非常荣幸能够应邀参加这次艾特奖的颁奖典礼，对艾特奖发展至今的成就也很钦佩，它已经具有重要的国际影响力。艾特奖国际学术委员会，可以在思想的交流、理论的发展方面产生作用，让我们的设计师有更深层次的思考，我觉得会对中国的设计发展会起非常重要的作用。

I am honored to be invited to attend the Awarding Ceremony of Idea-Tops. I admire the current achievements made by Idea-Tops. It has generated an important international influence. Idea-Tops International Design Forum plays a role in exchange of ideas and development of theory. It makes our designers think deeper and I believe it will play a very important part in Chinese design development.

刘育东
哈佛大学建筑设计博士
亚洲大学副校长

Liu Yu-Tung, Doctor of Design in Harvard University, Vice President of Asia University, Taiwan

艾特奖从名字来看就是Idea Tops就是点子最重要。所以世界各国的点子都能够汇集在一个平台上，一起被交流，被看到，我认为这就是艾特奖最重要的价值。它并不是在单一的文化、单一的设计价值观理念、单一的生活风格里面。而是能够跨越这么多的，不同的文化、气候、风土人情，以及不同的社会。

"By literal meaning, Idea-Tops is all about the ideas. Ideas in worldwide scope gathering on the same platform, to be appreciated and studied, it is what Idea-Tops matters in my point of view, which combines different cultural background, value and concept, weather, tradition, economy and life style. For that reason, it makes Idea-Tops the most unmissable event to be involved."

龚书章
台湾交通大学建筑研究所所长
建筑系教授

Gong Shuzhang, Director of Institute of Architecture, National Chiao Tung University, Professor of the Architecture Department

我们很荣幸艾特奖能作为一个重要平台，展现出了这个时代的经典作品。艾特奖在这个时候扮演了这个角色，让我们能够和其他所有好的设计师共同寻找那道光，我觉得这非常重要。

It's our great honor that the award—Idea-Tops has become an important platform for all typical design works in this era. More importantly, the award plays a role in standing together with all designers, especially excellent designers, in pursuit of excellence.

许楗
西安交通大学大遗址保护与
古建筑（国际）研究中心主任

Xu Jian, Director of Great Site Protection and Ancient Architecture (International) Research Center, Xi'an Jiaotong University

艾特奖举办了这么多年已经形成了非常大的影响力。今年更是一种转型，从室内为主转向了更广阔的领域，这很符合整个国际设计趋势和潮流，因为设计是互通的，对我们的室内设计师来说它能更加开阔视野，提升专业素质，提升应对目前整个发展的需求。

Idea-Tops has been held for many years and generated a huge influence in the world. This year, it has transferred from interior design to wider fields, I think that this complies with the overall international design trend and tide. Because design is interconnected, interior designers will widen their visions, enhance professional quality and cope with the demand of the overall development at present.

鲁晓波

清华大学美术学院院长、博士生导师
Lu Xiaobo, President and Doctoral Supervisor of Academy of Arts & Design, Tsinghua University

艾特奖国际学术委员会加强我们这个行业文化和知识底蕴的积累，这是一个国际性的学术交流平台，既是交流又是合作的平台。本届艾特奖参赛作品，相对来讲具有创新意义的前沿设计概念，或者是设计解决方案。我期待下一届会有更多更具特色、原创性的作品出现。

Idea-Tops International Design Forum has strengthened the accumulation of cultural and knowledge deposit in this industry. This is a platform for international academic exchange and cooperation. Relatively speaking, the entries of this year contain innovative and cutting-edge design concepts or design solutions. I expect that more special and original works will appear in the next competition.

张月

清华大学美术学院教授、硕士生导师、米兰世博会中国馆项目执行总监
Zhang Yue, Professor and Master Tutor of Academy of Arts & Design, Tsinghua University, Executive Director of China Pavilion at Milan Expo

艾特奖是一个比较独特的平台，横跨了两个领域，既有大量的中国设计师参与，同时也有很多国外设计师参与。这是一个可以把两种不同文化共融，同时也可以做设计的比较和竞争的平台。

Idea-Tops is a unique platform which involves in two fields. A lot of Chinese and overseas designers participate in it. It is a platform for combination of two different cultures, meanwhile, a stage for comparison and competition of design.

Rosetta Sarah Elkin

哈佛大学景观建筑副教授、设计研究生院联合主任
Rosetta Sarah Elkin, Assistant Professor of Landscape Architecture, Co-Director of Risk + Resilience Master of Design Studies in Harvard University Graduate School of Design

非常荣幸来到这里，而且很高兴了解到艾特奖的更多信息。两天的会议中见到很多有趣的学者，他们有着不同的工作方式、规模和形式，就室内设计来说，艾特奖从其他学科中吸取灵感这一点很明智。我们需要更多地进行合作,需要有多方位、不同规模、多学科的团队。

I am very honored to be here. I was pleasant to finding out more about the Idea-Tops conference. It does seem that in the conference over the 2 days you have brought together very interesting scholars with different approaches to work at different scales and different formats, and I think that as an interior design focus you are very wise to take inspiration from outside disciplines, outside design disciplines. So we need to work together more and

Francois Penz

剑桥大学建筑学院马丁建筑与城市研究中心主任、都市环境/建筑教授
Francois Penz, Director of The Martin Centre for Architectural and Urban Studies in Faculty of Architecture, University of Cambridge

艾特奖聚集了来自中国各地的人，以及世界各地的设计师、教授和重量级人物。我觉得互相学习和交流对现在和将来都大有益处。在这些奖项中，我们期待看到的是连结过去并展望未来。我觉得这些年轻的中国设计师一定能为设计界注入活力并带来新的想法。

Idea-Topsbrings people from all over China. It brings also designers, professors, and VIPs from all over the world. I think that we will benefit to learn from each other, to communicate with each other for the present but also for the future. For those awards, what we are looking forward is consolidating the past but also looking at the future. I

David Howard

牛津大学城市可持续发展学系主任、教授

David Howard, Director and Associate Professor in Sustainable Urban Development at University of Oxford

我觉得艾特奖很了不起，评审小组需要对一系列的类别进行评审，这些类别很不错，有很多优秀的参赛作品，有国际化的应用，我觉得评审小组非常乐于探讨这些很优秀的原创作品。最优秀的作品不仅展示了强大的技术能力，也展示了独创性。

I am very impressed by the Idea-Tops awards and the review panel asked to review a series of categories. And they have been very amazing categories, many excellent entries, good international range of applications, and I think the review panel very much enjoys discussing many good and many original entries. So the top entries, the ones that shows great technical ability, but also originality.

Antonino Saggio

罗马大学建筑学院建筑和信息技术系主任、教授

Antonino Saggio, Dean and Professor of Architecture and Information Technology Department at the School of Architecture, University of Rome

我很高兴来到艾特奖的颁奖盛典，我作为评审团的一份子，评选出室内设计、电子设计和建筑设计方面的最佳作品。艾特奖颁奖盛典的气氛与讲座都很不错，是一个值得期待的大奖。

I'm delighted to attend the Idea-Tops Awarding Ceremony. As a member of the jury, I selected the best works in interior design, electronic design and architecture fields. Atmosphere of the ceremony and lecture is very good, Idea-Tops is worthy of expectation.

Paul Lewis

普林斯敦大学建筑学院副院长

Paul Lewis, Associate Dean at Princeton University School of Architecture

我觉得作为设计师、建筑师、思考者，交流是非常重要的，照片和图像总是需要传达思想以及事物背后的想法。我认为这次盛会重要的一点就是将图片与思想联系起来。我非常高兴在这里能通过对话，听取到世界各地的人关于推动当今设计的各种问题和思维的意见。

I think it's incredibly important as designers, as architect, as thinkers that there is more dialogue that takes place so again questions of images, of photographs always need to be communicated to ideas and the ideas behind those things. I think one of the important aspect of today's event is to always connect questions of images to ideas. I'm so glad to hear different opinions from around the world about the questions and ideas that are driving

Yushuke Obuchi

东京大学建筑教授

Yusuke Obuchi, Associate Professor in Architecture at the University of Tokyo

以全球化标准来进行分享是件非常好的事情，但我觉得这展现了高端的文化、好的标准、好的质量以及我们是怎样定义设计的质量，这些或许更偏向于西方价值观。对我来说，这次盛会使得各种知识和信息在参与者以及更大范围的来自中国和世界各地的受众之间得以分享。

I think it's great to share in a global standard, as I'd say, I don't know if I should say this or not, but certain kind of high culture, good standard, good quality, how we, perhaps more western value I think, think what is the kind of quality of the design. It seems to me that the event allows those knowledge and information to be shared among the people who participate in but also wider

Takehiko Nagakura

麻省理工学院建筑教授
Takehiko Nagakura, Professor at Massachusetts Institute of Technology

艾特奖有趣的地方就是它很广泛，不限于某种特定的建筑，大体上属于室内设计范畴但又包含一些建筑成份。我认为这是关于世界现在所发生的事情的总概括，同时又像是不同学科的智慧的交融。我们都来自建筑设计这个领域。我认为这是当代建筑研究的交汇点，是建筑体验的前线。

The interesting thing about this award is it's very broad. It's not limited to certain kind of architecture but it's very broadly about interior design with a little bit about architecture components. I think it's a great overview of what's going on in the world. It looks like also a great mix of intelligence from different disciplines. We are all in the area of architecture design. So I think it's a great cross-section of contemporary architecture research and architecture front line of experimentations.

Martin Bouchier

巴黎高等建筑学院教授
Martine Bouchier, Professor of The Superior National School of Architecture Paris-Val-de-Seine

我非常荣幸受邀来到本次艾特奖。对我来说这是一个机会，能让我见到负责不同项目的伙伴，我以前并不太了解中国的设计。作为评审团的一员，我有机会见识到很多项目，我感觉所有项目操作性都非常好，非常有创意，项目中不同材料的运用令我惊叹。

I was honored to be invited to Idea-Tops meeting. For me it was an opportunity to see partners from different projects, I was not used to know your Chinese design. As a member of the jury, I have the opportunity to see many projects and my feeling is that all these projects have great quality of implementation. And the project is really creative, the uses of different material are very amazing to me.

Mihkel Tüür

2015米兰世博会
爱沙尼亚馆主设计师
Kadarik Tüür, Chief Designer of the Estonian Pavilion at Milan Expo 2015

艾特奖是一场很棒的盛会，组织也很好。能够和来自不同机构，不同国家的人分享自己的想法是很有意思的，我们得以分享知识，形成跨文化的互动。

Idea-Tops is a nice event, nicely organized. It's quite interesting to share ideas with different people from different institutions and countries, and we can share the knowledge, and to introduce it as a multi-culture.

Klaus K. Loenhart

2015米兰世博会
奥地利馆主设计师
Klaus K. Loenhart, Chief Designer of the Austria Pavilion at Milan Expo 2015

我相信这次与全球学术界及设计界的人士进行更紧密的沟通对话，是一个很重要的举措。艾特奖更加开放地面向整个世界，这是非常重要而且成功的一步，它通过全球在设计方面可能共同关注的问题创造了一种更紧密的联系。

I believe it is a very important initiative to start a much closer communication and discourse with other areas of academics and design on this plant. Now Idea-Tops opens more up to the world-wide community, I think it's very important and a very successful step to get into closer relationship with probably the same concern we have all over this planet in terms of design.

Idea-Tops Awarding Ceremony 2015
2015年度艾特奖颁奖盛典

2015年12月3日，深圳市政府领导、全球顶尖学府的设计专家教授、米兰世博会国家馆设计师代表、35个不同国家和地区的优秀设计师、艾特奖全国分赛区代表团、材料商以及设计企业高管、房地产龙头企业负责人、新闻媒体记者等近千人齐聚"设计之都"深圳，共同见证2015年国际空间设计大奖Idea-Tops艾特奖全球17位设计骄子的诞生。

作为目前中国国际化程度最高的建筑室内设计类奖项，本届艾特奖共收到来自美国、意大利、德国、荷兰、葡萄牙、比利时、日本、澳大利亚、西班牙、奥地利、希腊、俄罗斯、墨西哥、香港、台湾、印度、罗马尼亚、以色列、巴西等35个国家和地区的建筑与室内设计师参赛设计作品5682件，除了提交作品的数量和质量创历届之最，2015米兰世博会11家国家馆的参与，进一步凸显了艾特奖的国际影响力。

香港卫视、深圳卫视、南方电视台、广东电视台、凤凰网、中国网、中国新闻社、人民网、《南方都市报》、《南方日报》、《深圳特区报》、《深圳商报》、《广州日报》等近百家主流媒体聚焦本届颁奖盛典，在社会各界引起广泛关注和强烈反响。

On December 3rd, thousands of people gathered in Shenzhen, the City of Design, and witnessed the birth of 17 global talented winners of the International Space Design Award—Idea-Tops in 2015. Representatives of Shenzhen government, design professors from top universities in the world, representative designers of national pavilions from Milan Expo, excellent designers from 35 different countries and districts, representatives of Idea-Tops divisions, material merchants as well as senior managers of design companies, leading real estate company principals, news organizations and journalists attended the activity.

As the most internationalized award in the architecture and interior design filed in China, Idea-Tops has received 5,682 entries from 35 countries and regions in 2015, such as America, Italy, Germany, the Netherlands, Portugal, Belgium, Japan, Australia, Spain, Austria, Greece, Russia, Mexico, Hong Kong, Taiwan, India, Romania, Israel, Brazil, etc. The quantity and quality of the competition works are the highest in history. The participation of 11 national pavilions from 2015 Milan World Expo further highlighted Idea-Tops' international influence.

About a hundred of mainstream media broadcasted this grand ceremony, including HKSTV, SZTV, TVS Television, GDTV, Phoenix TV, China.org.cn, China News Service, People.cn, Southern Metropolis Daily, Southern Daily, Shenzhen Special Zone Daily, Shenzhen Economic Daily and Guangzhou Daily. It had attracted great attention and wide interests from different communities.

Idea-Tops 艾特奖

2015 INTERNATIONAL SPACE DESIGN AWARD
IDEA-TOPS 艾特奖颁奖盛典
AWARDING CEREMONY

026

IDEA-TOPS
艾特奖

028 | IDEA-TOPS
艾特奖

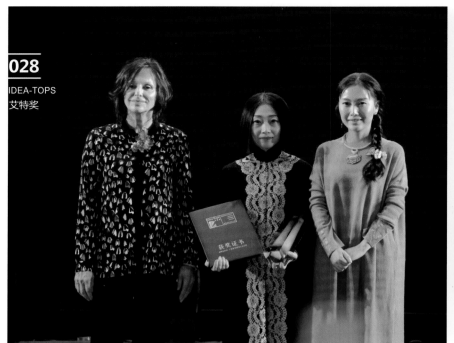

029
IDEA-TOPS
艾特奖

030
IDEA-TOPS
艾特奖

031
IDEA-TOPS
艾特奖

The 2ⁿᵈ G10 Designers' Summit in 2015

第二届G10设计师峰会

2015年12月2日,第二届G10设计师峰会在深圳市龙岗区中海凯骊酒店隆重举办。汇聚三十城力量,百位设计名家以"未来设计探索"为主题,深度把脉行业现状,共启未来新思路、新方向。

G10设计师峰会,由Idea-Tops艾特奖组委会倡导并发起,定位每年一届,于国际空间设计大奖——艾特奖颁奖盛典期间举行。G10(英GROUP10),是10个研讨小组的简称。为达到深度探讨交流之目的,每届G10设计师峰会限邀100位不同地区、不同文化背景且极具代表性的设计师参与,旨在为每一位参会设计师提供一个平等的发声机会、思想碰撞以及交流合作的平台,以此引领设计业的发展,推动世界设计产业的共同进步。依托中国境内国际化程度最高的设计大奖——艾特奖,G10设计师峰会备受业界关注。

2015年,建筑及室内设计行业面临更多挑战与机遇,"互联网+""一体化家装"等热点此起彼伏,重组并购也成为各地设计机构探索发展方向的重要选择。在成功举办2014首届G10设计师峰会的基础上,2015年第二届G10设计师峰会特邀国际著名设计师、香港PAL设计事务所创办人及首席设计师梁景华博士担纲峰会主席,并以"未来设计探索"为主题,齐聚来自中国华北、东北、华中、华南、香港、澳门、台湾及欧美地区等30个城市与地区的业界精英,就当前业界十大热门话题展开深入探讨。

On December 2nd, The 2ⁿᵈ G10 Designers' Summit was grandly held at Coli Hotel, Longgang District, Shenzhen. With the theme of "exploration of future design", 100 famous designers from 30 cities came together to analyze the industrial status and open new ideas and direction of the future.

G10 Designers' Summit was advocated and initiated by Idea-Tops Organizing Committee. It is annually held during the Awarding Ceremony of International Space Design Awards—Idea-Tops. G10 (Group 10) refers to 10 discussion groups. To achieve deep discussion and communication, 100 representative designers from different regions and cultural backgrounds will be invited to attend the summit, aiming to provide every attendee with an equal expression opportunity, a platform for communication and cooperation so as to lead the industrial development and promote the common progress of the global design industry. Being supported by the most internationalized design award in China—Idea-Tops, the G10 Designers' Summit has obtained tremendous attention from the design community.

In 2015, the industry faced more challenges and opportunities. "Internet +", "integrated home decoration" and other hot topics emerged one after another. Reorganization and merger have become an important choice for various design institutions to explore their development direction. The first G10 Designers' Summit held in 2014 was a success, and Idea-Tops Organizing Committee invited Patrick Leung (a world-known designer, founder and chief designer of PAL Design Consultants Ltd.) to serve as the president of the summit. Taking "exploration of future design" as the theme, the summit gathered elites from 30 cities and districts, such as North China, Northeast China, Central China, South China, Hong Kong, Macao, Taiwan, Europe and America, to discuss 10 hot topics in the current industry.

IDEA-TOPS
艾特奖

036
IDEA-TOPS
艾特奖

038
IDEA-TOPS
艾特奖

039
IDEA-TOPS
艾特奖

Idea-Tops International Design Forum
艾特奖国际学术委员会

2015年12月2日，艾特奖国际学术委员会成立大会暨首届学术会议在深圳市召开。

艾特奖国际学术委员会由艾特奖组委会携手全球22所顶尖级大学(包括哈佛、耶鲁、剑桥、普林斯顿、麻省理工学院、清华、北大等)的知名教授及学术领军人物发起成立，由中国科学院院士郑时龄担任学术委员会主席，哈佛大学建筑学博士、台湾亚洲大学副校长刘育东担任副主席兼执行主席，清华大学美术学院院长鲁晓波担任副主席。

国际学术委员会的正式成立，为艾特奖建立了强大的学术支撑，使得艾特奖站在新的学术高度，以更高的国际视野促进中西方设计的文化交流及学术传播。正如中国科学院院士郑时龄所说："艾特奖国际学术委员会的成立是为了进一步推动全球化时代中，设计界的国际交流、合作与发展；国际学术委员会的成立对我们提出了新的要求，激励我们的工作，要求我们按照国际学术委员会的目标，推动社会创新，扩展信息教育，发挥艾特奖在设计领域的专业影响力，拓展设计的跨界发展和努力。"

On December 2nd, 2015, the Establishment Conference of Idea-Tops International Design Forum & The First Academic Conference was held at Shenzhen.

Idea-Tops International Design Forum was established by Idea-Tops Organizing Committee, well-known professors and academic leaders from 22 top universities in the world, including Harvard, Yale, Cambridge, Princeton, MIT, Tsinghua University and Peking University. Zheng Shiling (Academician of the Chinese Academy of Sciences) serves as chairman of Idea-Tops International Design Forum, Liu Yu-Tung (Doctor of Design at Harvard University, Vice President of Asia University, Taiwan) serves as vice chairman and executive chairman, and Lu Xiaobo (President of Academy of Arts & Design, Tsinghua University) serves as vice chairman.

The establishment of Idea-Tops International Design Forum has provided strong academic support to Idea-Tops, and enabled Idea-Tops to promote cultural exchange and academic communication between Chinese and western design at a new academic level and with a higher international vision. As Zheng Shiling said, "the establishment of Idea-Tops International Design Forum is to further promote the international communication, cooperation and development in design circle in the era of globalization. Its establishment imposes new requirements on us and motivates our work. According to its objective, it requests us to promote social innovation, expand information education and make efforts on crossover development through Idea-Tops' professional influence in the design field."

043
IDEA-TOPS
艾特奖

The 2015 Idea-Tops International Master Forum Shortened the Distance between China and the World

2015艾特奖国际大师论坛拉近中国与世界的距离

2015年12月3日,作为2015年艾特奖颁奖典礼的核心内容之一,国际大师论坛以"21世纪的设计探索"为主题同期呈现。

依托艾特奖国际学术委员会的强大学术平台,2015艾特奖国际大师论坛盛邀包括首批艾特奖国际学术委员在内的著名学者发表精彩演讲,并由哈佛大学建筑学博士、台湾亚洲大学副校长、艾特奖国际学术委员会副主席兼执行主席刘育东担纲主持。此外,2015米兰世博会8大国家馆主设计师重磅同台,众彩纷呈。艾特奖国际大师论坛带来关于世纪设计探索的顶级思考,再次拉近中国设计与世界的距离。

As one of the core activities in the 2015 Idea-Tops Awarding Ceremony, the International Master Forum was held on December 3rd with a theme of "Design Exploration in the 21st Century".

With the support of the strong academic platform of Idea-Tops International Design Forum, the 2015 Idea-Tops International Master Forum invited famous scholars to make speeches, including members of the Idea-Tops International Design Forum, and appointed Liu Yu-Tung, Doctor of Design in Harvard University, Vice President of Asia University, Taiwan and Vice Chairman and Executive Chairman of Idea-Tops International Design Forum as the host. In addition, chief designers of eight national pavilions at Milan Expo 2015 shared a stage and contributed to a brilliant forum. The forum stimulated deep thinking about the design exploration of the century and shortened the distance between China and the world.

艾特奖

最佳别墅设计奖
BEST DESIGN AWARD OF VILLA

IDEA-TOPS

INTERNATIONAL SPACE DESIGN AWARD

051
IDEA-TOPS
艾特奖

获奖者/ The Winners
陳相妤&楊岱融
（中国台湾）

获奖项目/Winning Project
光井/ Lighting well

052

IDEA-TOPS
艾特奖

获奖项目/Winning Project

光井
Light well

设计说明/ Design Illustration

街屋式建筑是台湾最常见的住屋型态，但长久以来都有因房屋紧邻并排，故窗户只能设置于前后的问题，造成采光及通风不易。台湾属海岛型气候，房舍容易因空间的限制使屋内较为阴暗潮湿，本案为此类建筑，加上基地坐向属东西向，开口面西，且占地面积只有60m²地坪狭小，难以成为大多数人选择居住的对象。

设计师尝试为此类老式住宅创造出难得的采光与通风，将空间活化的最大值完美呈现。本案实际坪数仅有240m²，如何在有限的空间中取舍，成功兼具完整的住宅机能与工作室需求，实现SOHO族梦寐以求的完美空间，成为本案设计的核心目标。

概念以"穿透"为主题，将阳光、空气巧妙融入空间成为建筑的一部分。天井以强化胶合玻璃作为楼层间的连结，光线得以自屋顶流泻而下，使住宅的每一个角落都能拥有舒适的采光，贯穿整栋住宅的透明步道，无扶手且踏阶透空的黑色钢板楼梯可产生宽阔的空间感。

在每层楼面只有60m²的格局中，为了在有限的空间兼顾多重机能，将车库右侧地板架高，使用大片玻璃建材将玄关隔出，看似将门口空地规划为车库的传统做法，但实际上将空间作了完整区隔，塑造出独立的入口廊道与下方收纳空间。自玄关沿着黑色钢骨楼梯而上，即可进入主接待区，开放的空间既是家庭的餐厨区，也是同时兼具客厅及会议室，主要空间摆放了原木长桌，让原本冷冽的空间变为柔和。在工作室与主接待区之间使用双面柜收纳的方式，取代只用墙面做隔间的一般做法，争取更宽敞的空间及完善的机能。

在私人空间的设计上，为使三楼主卧室空间更为宽敞，舍弃了浴厕隔间，透过洗手台与梳妆台结合的平台划分区域，利用水泥铺面与木地板铺面划分出空间，以线的概念取代原本面的做法，浴室与厕所并共享拉门、收纳空间隐藏于墙面，看似同属一个空间的范围，却依然保有各自独立的机能。

顶楼成为调节整栋建筑的空气层，而透明天井成为整栋楼的光源入口，户外的草皮不但有降温的功能，也在居家及工作中加上了休憩的元素，搭配各层楼梯间所采取的无阻隔作法，让来自于顶楼的阳光及空气完整的洒落，每一处的设计意皆是对应居住需求的要素，使整栋楼的温度、湿度与采光皆透过多重的规划得到了最佳的平衡。

利用老屋原有的内部格局，在建材及工法上取得更大的空间感，创造了居住上梦寐以求的明亮采光与开阔动线，实现了生活与工作既可区隔，也互相契合的理想生活空间。

机能空间：一楼为大门玄关和车库，利用架高地板和玻璃隔屏创造出各自独立的区块，增加收纳机能。

少即是多：黑色钢骨阶梯为贯穿全栋的主要架构，减少扶手仅保留简约线条，无论视觉上或实际通风与采亮度都大幅增加通透感。

通透采光：每层楼皆设置了透明天井步道，将阳光自屋顶带入屋内各个角落，兼具实用与美观，并解决了屋内中段阴暗的问题。

空间的连通：透过透明天井与中空楼梯，打破了每一层楼之间的距离，房间、餐厅、车库，使生活的每个空间相互连结。

空间整合：运用交集的方式改善空间不足的问题，整合走道空间，删去楼梯扶手，墙柜造型平整化，崁入铁件柜体，营造空间放大开阔的效果。

打破空间隔阂：主卧采用全开放式设计，删去了浴室隔间，睡床与浴室共享走道，以浴洗手台和梳妆台作为房内区块的分隔，并将收纳空间隐藏于墙面，小坪数房间亦能拥有完整机能。

动线机能：台面延伸至墙面，将包含镜面和收纳的橱柜打开即成为梳妆台，浴厕使用共享玻璃拉门，同一动线具有多种功能以减少动线的浪费。

环保绿意：四楼客厅西侧为户外花园，可为二、三楼调节西晒温度，并增加住家的生命力，使人与空间和环境融为一体。

流通空间：辽阔通透的空间可随时感受到对方的存在，让人与空间、阳光与空气都能够不受楼层的阻隔穿梭在屋内。

Townhouse is the most common housing pattern in Taiwan, since the side-by-side position, the window only available at the front and backs which generally causing the problem of unventilated, humid and dark inside. This case in one of the pattern, andit faced west and with arrow floor size, which combine all the adverse conditions.

The designers provide several solutions for Taiwanese traditional apartment as this case to revise the prejudice. Through creative design, townhouse is also having the possibility for full lightening and ventilation. With only 240 square meters of total floor area, the dream house of overall living and studiofunction comes true.

"Penetrability" is the main concept for the house: For sunlight, air, water and people go through all the space smoothly. The original stairwell reconstructed to transparent patio trail and the black SRC stairs without handrail. The patio use glassto link the aisle, and for the sunlight shines down to every corner. We use the SRC stair instead of traditional brick stairs, the non-handrail and pierced design provide more space and a path for circulation, that both link and given spacious for each floor.

Each stair only have 60 square meters, the designeradopt subtraction andintersection solution to multiply the function of this small space. The right side of garage is the glass wall lobby that isolates the automobile exhaust and creates an individual entrance and storage. Going up with the SRC stairs is the main meeting space, the open kitchen-dining area is also use as meeting room, and the wood made dining table soften the black and white color. Between the meeting space and studio, we did not use compartment but a double side storage contain all the kitchen appliance, studio printers, and also lines and wires. We reduce the waste of space and provide a wider and spacious meeting and working area with functional design.

As for the private area, there are even lass compartment to integrate the pathway. We design a no-door bathroom for the mater bedroom at the 3rd floor, only zoning the sleeping and bathing area by the platform of washbasin and dresser table and different material of the floor. The shower and toilet share a slide door, hide the storage in the wall, by the concept of reuse for space, the boundary of area are less clear-cut but still able to keep their own function, and create a wide and comfortable living space. The living room on the 4th floor is the air zone of the whole building. The transparent patio brings in the sunlight, the penetrate stairs open the air convection, and the garden at west side adjust and lower the heat for 2nd and 3rd floor. Everyone of these design is echo the key element for living. The multiple function of 4th floor are the best solution contribute to benefit the temperature, humidity and lightening for the house.

By substantiallymodified the interior of the traditional apartment, we lead sun and air to every space in the house only from the rooftop, and initiate a dream house of great sunlight andcapacious. It is the idea living space while daily life and working area are independent yet connected.

Multi-functional
The 1st floor is the main entrance and garage, by using the glass wall and 架高地板 to separate the using of space, and not only gain more storage but also isolate the entrance from car exhaustgas.

Less is more
The black SRC stair is the main structure of the house, deduct the handrail with only simple stair line create a visual fresh look and benefit the lightening and ventilation.

Penetration of light
There is the transparent patio trail each floor that brings the sunlight from the roof down to every floor. It is a design for practical usage and aesthetic, and a solution of dark in the middle of the apartment.

Connection between spaces
Through out the patio and stairs; bedroom, dining area, garage and the entrance, we link every living space together.

Space integration

Break the barrierof space
The open master bedroom breaking the compartment of sleeping and bathing area and share the same aisle, only use the washstand and dressing table as the invisible dividers. The storage space all hide behind the wall, for the small room to have all in 1 functions.

Flow and function of space
The washstand platform extended to the wall, while open the cabinet with mirror and cosmetic products storage, the space turn to be the dressing table; the shower and toilet share a slide door;all these design use use the path flow and deduct the waste of space.

Eco-friendly
The west side of 4th floor is the outdoor garden, it helps to adjust the temperature for 2nd and 3rd floor, and also add lively energy for the house for people, space and eco are connected.

Air in the space
Living in the wide and spacious house, the host and hostess are able to sense each other everywhere. The sunlight, air and people passing though the space without isolated by floors.

IDEA-TOPS
艾特奖

056

IDEA-TOPS
艾特奖

057
IDEA-TOPS
艾特奖

获奖评语
精巧空间中让光和生活融合为一。
The exquisite space blends light and life into one.

058
IDEA-TOPS
艾特奖

NOMINEE FOR BEST DESIGN AWARD OF VILLA
最佳别墅设计奖提名奖

李智翔（中国台湾）

获奖项目/Winning Project

自由平面
le plan libre

设计说明/ Design Illustration

建筑建立在一座6m高的基座口，设计以两道长约15m的白色长向水平窗带开窗，勾勒出建筑的主要线条；大面的玻璃开窗，除了建筑面向的景观面考虑外，也师法Mies van der Rohe对于"框架结构"和"玻璃"材质的表现。我们希望框构骨架的近乎透明，模糊室内外空间的定义，营造宽广的空间视野。

建筑体正面覆加延伸的正方体串连室内外空间，镜面不锈钢材质融合外在环境与建筑。平顶式的屋顶花园延续建筑体矩形块状的简洁利落，如装置艺术般的牛奶盒(洗涤槽用途)翻洒泄出一滩的青草，为现代主义洒下幽默的词汇。

奉行现代主义建筑的精神："自由平面"与"流动空间"，简单的立面与精致的材质，如莱姆石主墙、白色大理石背墙与垂直绿色植生墙，纵横交错的垂直水平立面建构出形体的简洁与纯粹。4m×4m的楼板开口，刻意露梁，成为空间最有力道的线条，一楼延伸至二楼的石材主墙则连接两楼层的关系。

旧建筑存留的柱体不刻意隐藏于墙体中，使承重柱与墙分离。以不锈钢与白色石材包覆让柱体与墙产生若即若离的关系，彼此存在又彼此彰显个性。二楼环绕着开口的回字动线，依次是开放式书房、两间小孩房与娱乐室，同样奉行建构。

准则：自由平面与流动空间。

车库地坪铺面是以不同比例与质感的蓝、黑、白三色地砖，水平向性的构筑成一幅立体画作，加上车库尽头的锥形天井自然光的进入，呈现现代主义线条与光线的纯粹美感。

As the architect Mies Van der Rohe proposes "Skin and bones architecture", we try to keep all structure expose withtwo long horizontal sliding windows. Except for providing the house a spacious view and ample sunlight, the exposed frame and glass blur the line between indoor and outdoor.

Postmodern
The cube extend from the facade is an intermediary space between indoor and outdoor. The stainless steel reflects the blue sky and merges the architecture with nature. On the roof garden, we designed asink in the shape of giant milk box, which is toppled and sprinkled a pool of grass.

Neat façade
Pursue the principal of Le Corbusier's 5 points of new architecture.Neat facade with nature texture of material is being used. With atrium in 4m*4m scale, we exposed the beam as a line ingreat strength to intersectthe structure.

Free plan
To continue the principal of the free plan, we tried through the separation of the load-bearing columns from the walls subdividing the space. The column which in stainless steel and carrara becomes the visual points in the space.Circulation in the second floor flows like a square, across two bedrooms, study room and game room.

Linear unit
The floor in garage is combined with three kinds of tile that is in different scale. In the end of the garage, we built a skylight on the top to present the linear beauty of modernism.

059
IDEA-TOPS
艾特奖

060
IDEA-TOPS
艾特奖

IDEA-TOPS
艾特奖

NOMINEE FOR BEST DESIGN AWARD OF VILLA
最佳别墅设计奖提名奖

Sa Aranha & Vasconcelos（葡萄牙）

获奖项目/Winning Project

摩登&现代
Modern & Darling

设计说明/ Design Illustration

本案是一个现代感十足的室内设计，地板和墙壁运用了白色，与门和过道的黑色形成对比，此外，空间中还不拘一格地运用了红色和蓝绿色。这些对比强烈的颜色无疑让装饰变得独具特色。
各类物品不对称地摆放，不同颜色的巧妙搭配，让这个居家空间展现出年轻的活力和幽默感。为数不多的几件艺术品，却因其夸张的比例创造出一个有趣的现代空间。

With a very contemporary interior design, supported by the contrast between the white floor and walls and black doors and passages, this house still wanted to live from the irreverence of the red and turquoise. Strong and contrasting colors which will definitely mark the decoration.
A home that reflects youthful spirit and sense of humor present in the disproportion of some pieces and in this remarkable fusion of colors.
An environment that has few pieces but whose exacerbated proportions help create a contemporary and fun space.

065
IDEA-TOPS
艾特奖

066

IDEA-TOPS
艾特奖

NOMINEE FOR BEST DESIGN AWARD OF VILLA
最佳别墅设计奖提名奖

深圳市昊泽空间设计有限公司
（中国深圳）

获奖项目/Winning Project

林隐——紫悦府D户型别墅
Forest Conceal Villa

设计说明/ Design Illustration

生活可不可以像画一样留白，画的留白可以让视线和思维延伸到无限远，家的留白可以让身体和精神无限的自由舒展。我们把一个个彼此封闭的空间打开，将室内外的界限模糊，让空间流动起来。身体的自由穿行，也许能带来思想上随性放逐吧。

May life has a white space as a painting? The white space of a painting can make people's sight and thinking extend infinitely. Likewise, the white space of life can make people physically and in a free and infinitely way. Let us open every closed space to the outside world and make it flow freely. May physical freedom make your thinking wander at will!

NOMINEE FOR BEST DESIGN AWARD OF VILLA
最佳别墅设计奖提名奖

彩韵室内设计有限公司
（中国台湾）

获奖项目/Winning Project

木石·双重奏
Duo Wood and Stone

设计说明/ Design Illustration

自由流动的光感，赋予空间无可取代的正能量，整体规划上善用复层楼面特色，逐一安排主题鲜明的生活、娱乐机能。一楼前面为宽敞车库，后段规划为雅致的起居厅，二楼则是视野开放、通透的客、餐厅。本案大量使用木、石类素材整合全宅色温，为空间凝聚浓郁的休闲自然感，同时也施展精湛的现代工艺，勾勒出生动的景深层次与细节美感，当造型量体、构图画面不断在此间交汇、延展，生活中的人文深度与探索趣味也随之而来。流畅动线、简洁清透的介质处理以及低调但不附和一时流行的优质素材搭配，完成居室必要洗炼风格和机能定义，同时藉由内外不受限的光景呼应，赋予空间稳定、精致的包容力，特别是放眼所见垂直与水平线条间，灵活交织的力与美，精心勾勒和谐比例，重现细腻无匹的现代工艺！

Light, freely flowing across space at all times brings irreplaceable positive energy. The good use of over-and-under duplex planning enables organized arrangements of distinct thematic living and entertaining functions. A spacious garage sits in the front on the ground floor in front of the parlor room. On the second floor, there are the spacious, open living and dining rooms for the family. Public and private spaces, activity, and quietude are separated and extended by means of verticality. A large amount of wood and stones are used to integrate the color temperature of the entire dwelling and embody a strong sense of leisure and nature. Modern craft is exquisitely employed to lively depict the depth of view at different levels and the details at each level. When style, structure and constructs incessantly interchange and extend, the depth of the humanities and the exploration of interest in daily life arise.

071

IDEA-TOPS
艾特奖

074 B
Best Design Award Of Hotels
最佳酒店设计奖

艾特奖
最佳酒店设计奖
BEST DESIGN AWARD OF HOTELS

IDEA TOPS
INTERNATIONAL SPACE DESIGN AWARD

075
IDEA-TOPS
艾特奖

获奖者/ The Winners
郭士豪&杨焕生
（中国台湾）

获奖项目/Winning Project
冠月·萃/ Moon Elite

076

IDEA-TOPS
艾特奖

获奖项目/Winning Project

冠月·萃
Moon Elite

设计说明/ Design Illustration

酒店设计如雨后春笋般兴起，有的传统酒店空间配置一二种风格设计完成一栋酒店设计，将空间细节慢慢琢磨于生活细节中；有些酒店富有创意，变换万千风格琳琅满目，任何形式呈现都是难免如同，就像那"千江有水千江月"一般自然。群山围绕的山城有一处埔里小镇，群山不高但潭却很美，是所有国内外旅游人士均会到访景点及歇息、交流之所在，冠月·萃藏身于小镇里，远离都会繁华，寻求一悠闲山居生活。

设计核心概念是和谐对比，以蓝色与黄色作为水面记忆点区分两种不同房型搭配黑白灰三色融合在空间之中，为了突破典型旅馆空间的重复与单调，将细小而不起眼的凹凸条砖经由排列及重组变成空间里重要的完美墙面，台湾特有和平白大理石也巧妙地拼出高尚且独特的纹理，融合在空间之中。点子无所不在，捻来即可使用，在地性的宗旨不同的比例、肌理、材料及空间，设计团队试图突破既有框架，创造属于埔里小镇在地旅馆空间。材料或许很简单也很便宜，随处可见可取。只要能与当地文化、自然连结有认同感就是空间中的上等材料，这样的信念是对埔里山城环境伦理的尊重，也是设计的初衷。

房型规划：[冠月·萃]房型设计为标准旅馆单元空间，规划出45m²及55m²两种房型，两种房型均依浴室开口方式及配置而有所变化，让标准单元空间也有其巧妙的变化性。

Hotels' designs sprang up like mushrooms, some method of design just like traditional space allocation, and use one or two types to set on all hotel rooms, and some method of designs are full of creative with myriads of various change types. As the saying goes "A thousand moons on a thousand rivers" nature, it's inevitable that any presentation are similar.

In middle of Taiwan, there is a place surrounded by mountains named Pu-Li, the mountains are not so high but Sun Moon lake is magnificent, as known by travelers, also a world famous place. This is what "Essence" at, away from the bustling city, seeking to live in a cozy life.

The main design concepts are the comparison and harmony, to let blue and yellow as the main colors to separate between difference types of room, and match with black, white, and gray these three basic colors. In order to break the monotony and repeated of typical hotel space, rearrange the monotonous bricks of wall into the attractive wall, and Taiwan's special marble also shows the noble and unique of texture through the arrangement

Ideas are all around, grab it and use it. Try to break the past thought, use them as new look. To create the features in location belonging the local hotels, from the scale, texture, material and space, perhaps the materials are simple and with fair price, or can be seen and used everywhere, but we could have the same identity that when local culture connect with local nature is a way to show our respect to the environment, also the original intention of design.

For room planning, The rooms design as the standard hotels' space, which are two types of planning 45 m² and 55 m². These two types are a bit different between the opening directions of bathroom and have some changes, to let standard hotels' space has more ingenuity variation.

078

IDEA-TOPS
艾特奖

079

IDEA-TOPS
艾特奖

获奖评语

本案通过大量的材料、颜色、光线和各种各样的环境提供了对空间的环球体验。

It offers a global experience of space through the great quantity of materials and colors, light and a diversity of environment.

081

IDEA-TOPS
艾特奖

NOMINEE FOR BEST DESIGN AWARD OF HOTELS
最佳酒店设计提名奖

上海牧桓室内设计装饰有限公司
（中国上海）

获奖项目/Winning Project

12号酒店
Hotel No.12

设计说明/ Design Illustration

宜春恒茂御泉谷国际度假山庄坐落在江西省靖安县御泉谷的上风上水之地，独特的地理环境和丰富的人文历史激发设计创作，设计的灵感源于陶渊明的《桃花源记》，创造出于自然融合的建筑群体，四周环绕着郁郁葱葱的树林和灌木，隐于山风之间。走入大堂，室内空间宽阔，天花鳞次栉比通向楼梯方向，墙面灰色石材雕刻，将桃花源记中的诗句映入客人眼帘，禅境古意扑面而来。镂空屏风隔墙的设计，让空间隔而不断，与具有传统韵味的家具和配饰相互交融，打断俗世的纷扰。柱子和天花上的装饰灯具，利用金属和木材两种截然不同的材料和质地，创造空间内在张力，透过锐利的金属、细腻的木质线条强调空间结构，精致笔挺的吊灯，令空间由内向外展现独特魅力。穿过悠长的走道，在灰白色空间配合温暖沉着木色格子窗棂，让客人逐步感受一种优雅的热情。天花的黑色线条设计，带有极强的导向性，墙面上特别设计的壁灯，与柱体结合，极大地丰富整个空间的层次感和趣味性。中式餐厅天花还原建筑屋顶原有结构，保持空间开阔，让客人在明媚的阳光和纯净的空气里自由呼吸，与周边环境自然融合。室内的家具，灯具，外观简洁质朴，宁静的色调营造出优雅的用餐气氛。其间的天花吊灯，巧妙引入毛笔元素，具有极强的后现代艺术气质。大包厢的墙面中式雕刻窗花的组合和书架造型，构建出一个轻松随意的环境。西餐厅的入口设计，采用中国传统的影壁灵感，采用铁艺镂空祥云图案设计，隐约透出的光线使整个墙面变

的朦胧虚幻，使客人还未进入，油然产生探访究竟的意愿。深灰色的调子，强烈的红色跳跃其间。沉稳平和的空间氛围利用戏剧的元素，进行抽象概括，形成一个中西和古今的融合。在酒店的客房设计中延续酒店整体的色调，将酒店新中式风格贯穿始终。卫生间刻意调整空间格局，铁艺玻璃国画隔断设计，可以让空间随心意自由变动与外界的视觉联系：开放连接或完全私密。客房主墙面以手绘花鸟国画展现细腻质地，顶部镂空花格，内灯光投射其上，飘逸出古朴浪漫的情调。在整个空间设计过程中，在各个空间、砖、石、席、麻等材质的一起使用，在各种材质质感的鲜明对比与变化中，提炼、概括，空间和时间模糊性表达，隐喻着对"世外桃源"意境表达所做的尝试。

Yichun Mont Aqua Resort is located Yuquan Valley, Jing'an County, Jiangxi Province, unique geographical environment and rich humanity history can motivate design creation. The design inspiration is originated from Peach Blossom Spring by Tao Yuanming. The building groups combining with the nature are created. It is surrounded with verdant woods and shrubs, hiding amid the mountain breeze.
In the hall, the indoor space is broad, the ceiling boards are placed closely side by side to the staircases, and the gray stones are carved on the wall. The verses in Peach Blossom Spring come
into eyes, implying deep meditation and ancient conception. The hollow screen partition is designed to separate the space, mingle with the traditional furniture and accessories and get rid of the earthly interruption. The decorative lamps on the pillar and ceiling are made with metal and timber to create the inner tension of the space. Sharp metal and exquisite timber lines stress the space structure. The exquisite pendant lamps make the space display the unique charm from the internal to the external.
Along the long corridor, the gray white space is matched with the warm and placid wooden lattice window to make guests experience the elegant passion. The black line of ceiling is of strong guidance. The specially designed wall lamps are combined with the pillar to enrich the layering and enjoyment in the space.

Chinese restaurant restores the original structure of the building roof to maintain a wide space. The guests breathe freely in bright sunshine and pure air and naturally combine with the surrounding environment. The indoor furniture and lamps are concise and plain in appearance, and the tranquil hue creates an elegant dining atmosphere. The ceiling lamps are skillfully introduced with the writing brush element to

083

IDEA-TOPS
艾特奖

show the powerful post-modern art temperament. The combination of Chinese carved window grilles on the wall and bookshelf structure in the big compartment create a relaxing and casual environment.

The entrance design of western restaurant is inspired by traditional Chinese screen wall. Blacksmith and hollow cloud pattern design is adopted to release the light which makes the whole wall misty and illusory. You will want to learn about something before entering the restaurant. Deep gray hue and strong red are added. Drama element is utilized in the stable and peaceful space atmosphere. It is abstracted and summarized to combine the Chinese and western, ancient and modern elements.

The design of guestrooms continues the overall hue of the hotel. The new Chinese style of the hotel is always considered. The space pattern is deliberately adjusted in the bathroom. The blacksmith glass with traditional Chinese painting is of separate design, making the space change freely and linking with external vision. It can be open for connection or fully private. The main wall of guestroom is painted with flowers and birds to display the exquisite texture. The hollow lattices on the top of wall. When the light shines over them, a simple and romantic sentiment is released.

In the process of space design, space, brick, stone, mat and hemp are used together. In the sharp contrast and changes of all material textures, the vague expression for space and time is abstracted and summarized to imply the artistic conception of "a fictitious land of peace".

084
IDEA-TOPS
艾特奖

085

IDEA-TOPS
艾特奖

NOMINEE FOR BEST DESIGN AWARD OF HOTELS
最佳酒店设计提名奖

深圳市同心同盟装饰设计有限公司
（中国深圳）

获奖项目/Winning Project

福建璟江大酒店
Fujian Jingjiang Hotel

设计说明/ Design Illustration

福建璟江大酒店是一家按照标准五星级酒店建造的豪华酒店，酒店座落于连江县丹凤路，酒店拥有客房322间套，4个大型宴会厅，可接待大型宴会和会议，豪华中餐包厢29个，酒店还设有全日制西餐厅、日韩料理餐厅、SPA会所、游泳池、健身房等齐全的设施，连江县目前仅有一家四星级饭店，尚无五星级饭店。福建璟江大酒店是香港维多利亚酒店管理公司管理的酒店之一。该项目总投资5亿元，总占地面积20多亩，楼高23层，是一家集特色餐饮、休闲娱乐、温泉保健和商务会议等功能于一体的综合性商务酒店。酒店建成后，已成为连江县城地标性建筑。

Fujian Jingjiang Hotel is a luxury hotel built to five-star hotel standard, located on Dangfeng Road, Lianjiang County. There are 322 guest rooms, 4 banquet halls for large-scaled banquets and meetings, 29luxury Chinese restaurant private rooms.Besides, we have all-day western restaurant, Japanese and Korean restaurant, SPA, swimming pool, gym and other full facilities. Currently, there is only one four-star hotel in Lianjiang County, not to mention a five-star which is under supervision of Hong Kong Vitoria Hotel Management Co.,Ltd. Five billion RMB has been invested into this project, whose area is over 1.33 hectares and stands 23 floors. Fujian Jingjiang Hotel is a comprehensive business hotel equipped with special dinning, entertainment, hot spring and business meeting service and facilities,thus it is now a landmark of Lianjiang County.

087
IDEA-TOPS
艾特奖

088
IDEA-TOPS
艾特奖

089
IDEA-TOPS
艾特奖

NOMINEE FOR BEST DESIGN AWARD OF HOTELS
最佳酒店设计提名奖

黄后钦（中国深圳）

获奖项目/Winning Project

温州莲云谷温泉度假
Lotus Spring Holiday Inn

设计说明/ Design Illustration

远离都市的繁杂与喧嚣，领略山谷之优美与宁静，是无数都市人内心的向往。
位于温州泰顺的莲云谷温泉度假酒店就是这样一处度假休闲胜地，酒店占地面积85亩，依山傍谷而建，拥有视野广阔的峡谷景观和得天独厚的温泉资源。酒店按五星级标准设计，整体呈简约的现代中式风格，加之对当地古民居及民俗文化的提炼融汇，营造出高端、奢侈的休闲度假氛围。
在接到该项目的设计需求之后，洲际设计团队在黄后钦先生的带领下对当地民俗、民居、少数民族文化、历史建筑文化等进行了深入全面的调研考察。项目所在地泰顺，被誉为"中国廊桥之乡"。"廊桥"，即为有屋檐的桥。一座座古朴的廊桥坐落于群山之间，成为独特的风景线，如长虹饮涧，新月出云。同时，泰顺也是畲族聚居之地，在畲族聚居的村落，人们仍然能体会到浓郁的上古遗风，让人如置身于世外桃源。基于休闲度假酒店的定位，从古朴的廊桥、丰富的民俗中撷取灵感，就地取材将木、竹、莲等元素运用于设计之中，设计师打造出了既蕴含当地特色文化，又具有现代舒适、奢华简约特点的酒店空间。
走进酒店大堂，回归自然、悠然假期的感觉扑面而来。传统特色的斜屋顶、背景墙上的莲花造型、中式风格的座椅、雕刻和木质吊灯，自整体到细节，风格统一又各具设计感，营造出的休闲氛围和古朴气质令人心情全然轻松。
除了美感与氛围的营造，设计师也特别考虑到了室内空间中的节能设计。电力成本在酒店的运营成本中占到非常大的比重，因此，如果通过设计能够节省电的成本，则可以大大降低其运营成本。该酒店地处山野，通风很好，设计师在设计的过程中充分利用这得天独厚的地理优势，结合室内外地势空间，让南北空间空气流通，使得夏天可以不用开空调，做到节能环保。
室内温泉休闲区根据不同的泡疗方式，设置了不同的体验空间。男女宾入口处以畲族文化里极具代表性的人物形象作为装饰，趣味十足。地下泡池区的一道廊桥别具风情，最初这里的原建筑只是仅有的几个柱子，设计师巧妙运用"廊桥"的概念将它们加以改造，成为空间的一大亮点。在这里，还能看到设计师原创的"房中房"，即将当地民居"移植"进来，在一个室内空间里又形成相对密闭的一排小房子，大大丰富了室内空间的立体感和趣味性。同样为空间增添了趣味性和想象力的，是设计师在天花上特意设计的蓝天白云，在天气寒冷的季节，当地室外温度可达零下十度，而在酒店温泉区里的客人，则可以感受到夏天般温暖的室外氛围，尽情享受酒店里的休闲时光。
酒店客房分为三种风格，根据建筑结构和空间大小等因素的不同，充分发挥空间的特质，满足空间的使用功能，为客人提供不同的居住体验。一楼客房利用空间本身的建筑感，更具休闲意味，房间内带独立泡池。房内墙面做肌理处理，地面铺砖以防潮。房间背景墙上的莲花元素，又将酒店的LOGO再次印入客人心中。在精致的日式风格房间里，同样以木质元素为主，但色调变为浅色系。小而精的客房对空间利用的要求更加高，设计师选用简约的家具，加之特色鲜明的装饰品，营造出温馨的居住氛围。
中餐厅的设计风格更加原生态，融入民俗，返璞归真。既有当地建筑文化中石柱的古朴，民俗文化里的竹文化、茶文化，也有最为经典著名的廊桥建筑，由此开启休闲度假之旅中非常重要的美食体验。除了当地文化氛围的萦绕，餐厅设计中的灯光对就餐体验的影响也非常重要。该空间的照明均采用聚照，餐台亮、其他区域暗，因对比而产生空间立体感，从而让客人感受到宁静、休闲的就餐氛围，且聚照到食物上的灯光能提升菜品观感，也会让客人更有品尝美食的好胃口。
不仅是餐厅，灯光对其他区域的氛围营造同样重要。客房、走道的灯光设计都不宜太亮，且必须是立体照明，不用泛光，而要用点光。如此灯光下的过道将产生立体感与进深感，给人以温馨浪漫的感觉。
与中餐厅相连接的酒吧空间，和中餐厅在功能设置上形成互补，同时也是各自独立的空间。鹅卵石装点的壁炉、树枝般肆意生长的围栏、石头铺就的地面、木质的斜屋顶、自然垂坠的枯草……构建成了这个原生态的酒吧空间，与家人朋友在此听音乐聊天，尤其在寒冷的冬日，当壁炉里生起木柴火，回归乡野、休闲享受的心境将由然而生。

Numerous city dwellers yearn for a life far away from noisy cities and able to appreciate the beauty and tranquility of valleys.
Lotus Hot Spring Resort Hotel, located in Taishun, Wenzhou, is such a leisure resort. Built by mountain and valley, the hotel covers a floor area of 5.67 hectares and boasts a broad gorge landscape and advantaged hot spring resources. Designed with the 5-star standard, the whole hotel features a concise modern Chinese style. By blending the local ancient dwellings and folk culture, a high-end and wild resort atmosphere is developed.
According to the design requirements, the design team of Intercontinental Design Company, led by Mr. Wang Houqin, investigates and surveys the local folk customs, dwellings, minority culture and historicalarchitectural culture, etc. The project is located in Taishun, known as "Hometown of Gallery Bridges of China". "Gallery bridge" means a bright with eave. Many simple and unsophisticated gallery bridges in the mountains form a unique landscape, just like rainbows over a river and new moon from clouds. Meanwhile, Taishun is a place where She people live. As there are still full-bodied customs of remote ages in the villages of She people, Taishun is like a land of idyllic beauty. According to the positioning of resort hotel, the designers are inspired by the gallery bridges and rich folk customs, use such elements as wood, bamboo and lotus, and create a hotel space featuring local cultural characteristics and comfortable, luxury and concise modern characteristics.
A sense of returning to nature and leisure will refresh you the moment you enter the hotel lobby. The traditional pitched roof, the lotus modeling on the background wall, Chinese-type chairs and wooden droplights, each with unique

characteristics as a whole and in details, create a leisure atmosphere and make you relax.
 In addition to aesthetic feeing and atmosphere, the designers also consider energy conservation design in interior space, in a way to lower operation cost for the hotel. Since the hotel is located in mountains with good ventilation, the designers make full use of such advantaged geographic strengths and make air flow from south to north, realizing energy and environmental protection because air-conditioning is usually not required in summer.
 Different experience spaces are set in the indoor hot spring leisure area according to different bath modes. The entrance is decorated with the representative figure of She culture, which is very interesting. A gallery bridge is uniquely designed as a highlight of the space by the designers by using the concept of "gallery bridge" for transformation. The houses of local residents are "transplanted" in a row by the designers, which greatly enrich the stereoscopic sensation and interest of the space. In order to enhance interest and imagination, blue sky and white cloud are designed on the ceiling. In this way, when outdoor temperature is minus ten degrees centigrade in winter, guests in the hot spring area of the hotel still can feel a warm outdoor atmosphere like summer and enjoy leisure time to their heart's content in the hotel.
The guest rooms of the hotel are designed with three styles to cater for different dwelling experiences of guests. The guest rooms on the first floor are designed in a more leisurely way, each with an independent bath pool. The wall surface resorts to texture processing, while the floor is tiled for damp proofing. On the background wall, the lotus element impresses guests of the LOGO of the hotel. In the exquisite Japanese-style rooms, wooden element still prevails, but light colors are adopted. Since the small and exquisite guest rooms have a higher demand on space utilization, the designers select concise furniture, coupled with unique ornaments, which creates a warm and cozy dwelling atmosphere.
 The design style of the Chinese restaurant is of more original ecological factors. It integrates folk customs, local stone pillars, bamboo culture and tea culture and classical gallery bridge construction, thus creating a very important catering experience in tourism. The lighting of this space adopts spotlights which lighten the dining table only, hence a sense of three-dimensionality is generated and guests can experience a tranquil and leisurely dining atmosphere. As the lights on foods can promote the impressions of dishes, guests will have a good appetite.
 Lamplight is also important for the atmospheres of other areas. The lamplight of guest rooms and corridors should not be too bright, while it should be three-dimensional point light. If a corridor generates stereoscopic sensation and spatialdepth sensation under the lamplight, it will give people a cozy and romantic feeling.

The bar space connected with the Chinese restaurant, in terms of function setting, is complementary to but isolated from the Chinese restaurant. A bar space of original ecological factors is created by the fireplace decorated with pebbles, the branch-like fence, the floor paved with stones, the wooden pitched roof, the naturally dropping withered grass…When listening to music and chat with your family members and friends, especially in a cold winter day, your desire for countryside, leisure and enjoyment will arise spontaneously with the wood fire of the fireplace.

IDEA-TOPS
艾特奖

094
Best Design Award Of Club
最佳会所设计奖

艾特奖

最佳会所设计奖
BEST DESIGN AWARD OF CLUB

INTERNATIONAL SPACE DESIGN AWARD

获奖者/ The Winners
李保华
（中国杭州）

获奖项目/Winning Project
中国景德镇云居草堂会所/Bamboo House

096

IDEA-TOPS
艾特奖

获奖项目/Winning Project

中国景德镇云居草堂会所
Bamboo House

设计说明/ Design Illustration

一个充满回忆的餐厅。
窗外的蝉鸣、院内的嬉闹、草间的萤光、姥爷的八哥、妈妈的红烧肉……儿时的记忆在这间餐厅里穿梭，整个餐厅看不到复古的物件和标签印迹，却充满浓浓的怀旧记忆的思绪。餐厅位于一条小巷弄，徽派民居的前庭后院格局被植入在这间沿街小店铺内，爬满植物的庭院和天井穿插其中，"吱咔"作响的隔栅屏风门投射的光影洒落在地上。增设的夹层扩大了使用面积，通过夹层预留中空使上下空间互动，将前庭后院的空间形态设置其中。
推门而入便是前庭，小鸟落在厨房的屋檐上，鸟笼里的帽子戏法在上演。
儿时嬉闹追逐的笑声在满是江南味的屏风门中穿梭。
院墙上的爬山虎伸到了窗前，荧光在草丛中闪烁，树上的风灯随风摆动。
拾阶而上，布满苔藓的后院充满了潮湿的味道。
房间里窗外的绿色乱入眼帘。在这里只有甜美的回忆。

ARestaurant Filledwith Memories.
Cicada sang outside the window, children chased around the courtyard, fireflies flashed among the grassland, grandpa's mynah and sauce meat cooked by mom…all these memories since childhood linger around this restaurant, even with no traceable sign of antique or restoring mark, a strong sense of nostalgic is endowed here. Located in a small lane, this restaurant adopts a typical Hui-style layout, implements front courtyard as well as a back one to this street-side space, walls crawled all over with plants connect patio, and the creaking folding screen creates a subtle play of light and shadow on the floor. The newly generated interlayer broadens usable area and improves the interactivity between spaces and subtly integrates the front yard and the back yard.
Stepping inside is the welcoming front yard, little birds resting on the roof of the kitchen playing the hat-trick in the cage.
Children chasing around, laughing echoes through the Jiangnan style folding screen.
Creeper crawl over the entire wall and extends twig ahead of the window, fireflies flash in the grassland, lanterns swing in the wind.
Climb up the stairs, the backyard is covered with moss and damp smell.
Bright green scene encompassed the world outside the window. All you can experience here are sweet memories.

获奖评语

该项目成功发明了应用传统材料,譬如竹子,来创建现代的休闲空间。
This project successfully re-invented use of traditional materials like bamboo in making of contemporary space for relaxation.

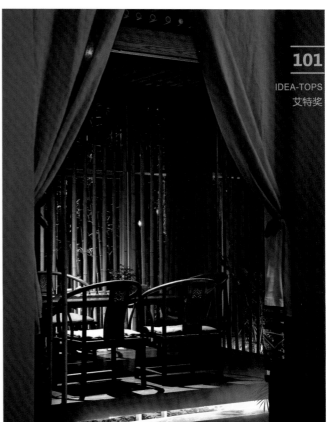

NOMINEE FOR BEST DESIGN AWARD OF CLUB
最佳会所设计奖提名奖

邵唯晏（中国台湾）

获奖项目/Winning Project

方寸间的皱褶
Imagination within Inches

设计说明/ Design Illustration

本案位于台湾中坜，业主是布料界的成功经营者，这是一个属于他个人的私人会所，一个招待朋友与偶尔自住的私人天地。整体的设计理念承载了业主对于美学的独到喜好和企业识别。布料是一种演艺性很高，充满生命力的材质，透过不同的外力会产生出皱褶，进而生成有机的肌理形变，方寸间演译出无限的可能。透过挤压、折迭起皱而成的线痕皱褶，会呈现出细碎锐利的褶面；而透过卷曲的起皱会产生柔和的曲面和边缘，千回百转的皱折创造了动态盎然的柔动姿态。凭借业主低调的艺术家性格，我们与他共同创造出属于他个人的小宇宙，在这他将最接近自己，一个属于自我的世界，一个专属的会所，一个自身的栖息场域，一个标志个人风格的奇景异境。

蜿蜒皱褶中的叙事网络

归纳我们对于布料的观察、体悟和想象，形式上从布料的皱褶出发，也隐喻了对于非线性形式的喜好与钻研，时间和空间随着物质本身的折迭、展开与扭曲，所形成的一种本质上没有内外之分的空间美学，这观念打破了欧几里德过往的几何空间概念，凝聚了一个动态运动中的片刻，我们透过有机、非线性、抽象的写意风格，创造了具有动态韵律、似地景、似装置、似墙体、似软装陈设的空间对象群，进而转译编织成一种超现实的诗意空间。

因而我们在空间中的许多角落都置入了这样展演性高的"空间对象"，散布在整栋建筑中，打破空间的主从关系，即使在最不重要的顶楼楼梯间角落，一样会觅寻到惊喜，生活的趣味就应该散布于整体的空间，透过单点对象的置放，串连后让空间充斥着叙事性的风格。甚至透过"隐门"的手法弱化了房间的自明性，从一楼一直到顶楼都在强调公共领域的空间，翻转了公私空间的定义，进入一个充满无限想象的艺术地景中，有如贤人雅士将奇山异水的景致收纳在皱褶的肌理中，柳暗花明又一村的空间安排，也将交织起属于这会所特属的叙事网络。

电视墙

经过大量的讨论，业主为了艺术同意牺牲了楼地板的面积，我们打开了二楼的楼板，创造出一个挑高8m的开放公共空间。在空间中置入了一个大尺度的空间对象，每日夕阳的余光透过云隙洒落在这块"布料"上，和皱褶肌理上演了一场光影秀，似在叙说着许多的故事，映射感染了整个空间。然而，除了结合电视墙的机能外，也企图藉此空间装置诉说着空间场域的精神，同时也承载了业主自身专业领域的企业隐喻。

沙发

位于一楼会客室的座椅设计也是量身订制，是一座充满动感有力度的曲面皱褶，在蜿蜒细碎的皱褶中找寻东方书法的柔情姿态，在沉静的会客室空间中恣意展现皱褶姿态，同时也加入了书法抛筋露骨、柔中带刚的线条，在具备了西方抽象艺术的现代表现基础上，也充满东方书法线条的动态语汇，期望用户在空间中凝神静思之时，品尝这交替运行所形成具有律动美的造型艺术。

数位的永续思维

整体空间因预算的考虑，也导入一些节能永续的思维，第一是整体空间的格栅都是搜集可再利用的实木角料经过漆料的修补所拼接构成，在制作复杂形体之余，将其剩余的材料转到其他空间所有利用。第二是空间中的复杂形体，在设计前期阶段，先在计算机参数化的环境中设计出转译出的形体，并利用快速成型技术(RP)输出实体模型来进行设计讨论与沟通，并在计算机中来回分析与修正。进入施作阶段，首先利用雷射切割制作成连续渐变的断面，并在曲面强度不同的地方进行结构的补强，之后开始精确放样，组立粗坯形体。从设计到施工的整个过程，都透过计算机辅助设计系统及计算机参数化设计流程精密控制，企图在压缩的预算及工期内，将无秩序的对象有效模矩化与制程化，经过精密的材料及形式计算，在工厂模块化完成后再运至现场组装，有效减短工期及节约现场施工资源。

This case is located in Zhongli City of Taiwan. The owner is a successful businessman in the cloth industry. The case is his personal private club for reception of friends and occasional self-occupation. The overall design concept manifests the owner's unique taste on aesthetics and the corporate identity. Cloth is a texture with high art performance and full of vitality. It generates folds under external force and further generates organic texture deformation. Infinite possibilities are deduced within inches. Via extrusion and folding, thread trails and folds are formed to present a finely sharp folding surface; while a soft curved surface and margins are generated through the curly folds and the folds of innumerable twists and turns create a dynamic and soft posture. According the artist character of the owner, we jointly create a private microcosm for him. Here, he will get close to himself and enjoy a world exclusive for himself. As a wonderful place with personal style, the property can be used as an exclusive club or for private inhabitation.

Narrative network in winding folds

By concluding our observation, understanding and imagination towards cloth, we metaphorically express our fondness and study of nonlinear form from the folds of cloth. Time and space form a kind of spatial aesthetics without difference in interior and exterior in essence along with the folding, unfolding and distortion of materials. Such a concept breaks the past geometric space concept of Euclid and agglomerates a while in dynamic motion. With the adoption of an organic, nonlinear and abstract artistic style, we create a dynamic and rhythmic spatial object cluster like landscape, device, wall or soft decoration furnishings, and then we translate and weave it into a surreal poetic space.

Therefore, we set "spatial objects" of high performance property in many corners of the space and distribute them in the space of the whole building, which breaks the master-slave relation of the space. Even though you are at the staircase corner of the top floor, you still can seek out surprise. The interests of life should be scattered in the whole space. When the objects placed at different points are connected, a narrative style will overwhelm the whole space. The technique of "jib door" weakens room brightness. From the first floor to the top floor, we emphasize the space of public domain, which overturns the definitions of public and private spaces. The space you will enter will be an artistic landscape full of infinite imagination. Just like sages hold the scenes of mountains and waters in crumpled texture, the enlightened spatial arrangement will interweave a narrative network exclusive to this club.

TV wall

After a great deal of discussion, the owner agrees to sacrifice the floorage for the sake of art. To create an open public space with a height of 8m, we open the slab of the second floor. A large spatial object is placed in the space. The glow of sunset sprinkles on this piece of "cloth" and performs a shadow show with the folding texture, just like telling a lot of stories and impressing the whole space. With the function of a TV wall, the spatial installation also aims at narrating the spirit of the field domain and carrying the corporate metaphor in the professional field of the owner.

Sofa

The design for the chairs in the reception room on the first floor is tailor-made. With dynamic folds, the designers seek the tenderness of oriental calligraphy and present an unscrupulous gesture in the quiet reception room space. Meanwhile, they add the thinly veiled, soft and firm lines of calligraphy. On the basis of the modern manifestation of western abstract art, the dynamic writing of eastern calligraphy lines is integrated. In this way, the user can enjoy the rhythmic and dynamic plastic arts when he contemplates in the space.

Sustainable thinking of digits

Energy-saving and sustainable thinking is introduced for budget of the whole space. Firstly, the gratings of the whole space are made by splicing the recyclable solid wood angle materials repaired by paint. After complex structures are fabricated, the residual materials will be transferred to other spaces for recycling. Secondly, the complex structures in the space are designed in a computer-parameterized environment in the early stage. After solid models are outputted by using RP technology, discussion and communication will be made for repeated analysis and amendment in a computer. After entering the construction stage, we firstly fabricate continuously varying sections by laser cutting and reinforce structures at the places with different curved surface strengths. Later, we start setting out accurately and assembly rough blank structures. In the whole process from design to construction, by fine control of the computer-aided design system and the computer-parameterized design process, we aim at making the chaotic objects modularly and processized within squeezed budget and project duration. Via precise material and form computation, assembly will be made after plant modularization, which effectively shortens the duration and saves on-site construction resources as well.

IDEA-TOPS
艾特奖

NOMINEE FOR BEST DESIGN AWARD OF CLUB
最佳会所设计奖提名奖

Georges Hung（中国香港）

获奖项目/Winning Project

宝贝会所
Babysteps

设计说明/ Design Illustration

宝贝会所位于香港中部亚毕诺道的"环球贸易中心大厦"25楼，这是新创建的学龄前托儿所，其利用创新心理的方法来刺激孩子对学习的热情，以培养一种独特的学习经历。围绕着游戏、音乐和艺术疗法，其重点是学习如何发展提高，同时确保每个孩子心理的健康成长。

在160m²的商业地板布局范围内，宝贝会所的室内设计围绕着3个主要理念：①亮度：开放、宽敞以及有魅力的布局；②勇气：增强简单性、空间深度而同时具有透明度和流动性；③颜色：以一种优雅和微妙的方式将色彩的喜悦带入到背景中。

室内布局围绕着主要游戏区进行组织，作为该中心的空间特性。它是一个开放、明亮、有魅力的游戏区域，同时具有多个区域的活动：活跃、触觉和平静。一面带有弯曲几何彩色图样的较大特色墙和内置的存储固定了空间，并提供了一个有趣的且相当于主要游戏区而不失平静的背景。这些图样从孩子们的水平以及成人水平，提供了各个教室的整体概况。在特色墙后面，有3个教室，与玻璃幕墙并排布置。这些教室是模块化的，因而在一天的不同时刻提供了亲密团体活动。在较大的教室和标准教室之间设置一个可移动的分区，使得一个较大教室的建立能够用于交互式数字媒体板。

入口区域的设计是为了迎接孩子们和八方来客。通过透明的玻璃门，该入口是由一系列弯曲的平面制成，采用垂直色彩和单色金属百叶窗进行组装。这种温和的波动展现了其自身作为接待区域以及包含4个工作站的员工区域。这些竖向条板的波动为宝贝会所创建了一个微妙的光影效果，并且精巧地制作了一种视觉识别效果。

储藏室和休息区是相互设计的两个区域，用于家长和老师进行互动。该储藏室是一个具有功能性布局的厨房，同时带有多种色彩的橱柜。它激起了人们的一种趣味感和异想天开的品质。这些色彩是由作为宝贝会所标志的相同颜色色调而演变过来的。它与该托儿所的视觉识别相结合，同时与室内空间的整个微妙的单色品质进行有趣的互动。

家长的"休息室"是一个休息区域，是由各种曲折组合平台制成的。整齐地塞在角落里，倚靠玻璃幕墙边缘，且不妨碍到孩子们的游戏区，该休息室让孩子们可以有一个完美的视角，就像一个很大且开阔用于生活和工作的阁楼空间。用于宝贝会所的内部装备是一系列有关联的教育功能，以一种流动性开放空间的精神进行布置，让孩子们同时感受到游戏和学习这样一个经历。

Situated on Arbuthnot road in Central, on the 25th floor of the Universal Trade Centre Tower, BabySteps is new playgroup created to foster a unique learning experience utilizing innovative psychological approaches to stimulate children's passion for learning. Centering around play, music and art therapy, the focus is on learning development while ensuring healthy psychological growth in every individual child.

Within the 160 sqm of commercial floor layout, Babysteps interior is designed around three main concepts: 1. Brightness: open, spacious and inviting layout. 2. Boldness: Enhanced simplicity, spatial

depth with transparency and fluidity. 3. Colour: Bringing the joy of colour in an elegant and subtle manner to the context.

The interior layout is organized around the principal playzone as the central spatial feature. It is an open, luminous and inviting play area with multiple zones of activities: active, tactile, and calm. A large feature wall with curved geometrical colorful cutouts and built-in storage anchor the space and provides a playful yet calm backdrop against the principal playzone. These cutouts provide glimpse into the classrooms from the children's level as well as adults level. Behind the feature wall, there are three classrooms that align the glass curtain wall. Modular, these classrooms provide intimate group activities at different times of the day. A moveable partition between the larger and standard classroom allows the creation of a larger classroom for the use of the interactive digital media-board.

The entry area is designed to welcome visitors and children alike. Through the transparent glass doors, the entry is made up of a series of curved planes assembled with vertical colour and monochromatic metallic louvers. The gentle undulation unfolds itself into the reception area as well as the staff area containing 4 workstations. The undulation of the vertical fins creates a subtle play of light and shadow and crafts a visual identity for BabySteps.

The pantry and the lounge area are the two zones mutually designed for parents and teachers to interact. The pantry is a functionally layout kitchen with multi-colour cabinetry. It evokes a sense of fun and whimsical quality. The colours are derived from the same colour tones as the Babysteps logo. It connects with the visual identity of the playgroup while playfully interacts with the overall subtle monochromatic qualities of the interior space.

The Lounge for the parents is a sitting area made up of zig-zagging built-up platforms. Tucked neatly in its corner and against the ledge of the glass curtain wall and without intruding onto the children's playzone, the lounge allows for a perfect viewing angle of the children.

Like a large open loft space for living and working, the interior fitout for Babysteps is a series of connected pedagogical functions laid out within the spirit of a fluid open space where play and learning intermingle into one experience.

NOMINEE FOR BEST DESIGN AWARD OF CLUB
最佳会所设计奖提名奖

Serrano Monjaraz Arquitectos（墨西哥）

获奖项目/Winning Project

Vidalta温泉会所
Therma Spa by Vidalta

设计说明/ Design Illustration

该项目在9层高以及和7000m²施工面积的俱乐部会所内的两层内完工了。各项服务就像世界一流的酒店那样，设施包括：餐厅、体育酒吧、健身房、室内娱乐台球和游泳池，以及游泳水槽、室外游泳池和水疗中心。

该空间的布局包括：双高度通道、休闲区域、按摩小屋、治疗小屋、零售商店区、淋浴和更衣间。而在另一端，在一个具有大窗户的18m高的室内空间里，感觉到池子有其自身三倍高度（只有在树木繁茂的峡谷里才能领略到的特权）。两个楼层采用一个圆柱形玻璃全景电梯加以连接。石材建筑结合了各种治疗方法，用于展现一个主题概念。桑拿、蒸汽房、淋浴庙、健康水疗中心、活力温泉以及sanotherm，使空间成为一个独特的经历，使得美学与最先进的健康护理相结合。该建筑以控制光线作为创建宁静氛围的一个元素，并且在一个单色空间里的给人以一种宁静般的感觉。

The project for Therma SPA by Vidalta was done in two levels inside the clubhouse that has 9 levels and 7,000 square meters of construction. The services provided are just like the ones given in a world class hotel and among the amenities are: restaurant, sports bar, gym, indoor fun pool and swimming pool with swimming flumes, outdoor pool and SPA.

The space was distributed as follows: double height access, relaxation areas, massage cabins, treatment cabins, retail shop area, showers and dressing rooms. On the opposite end the sensations pool with a triple height —that privileges the views of the wooded ravine— in an 18 meters high interior space with grand windows. The two levels are joined by a cylindrical glass panoramic lift.

The composition of the stone volumes unifies the various therapeutic uses in a thematic concept to experience, the sauna, steam room, the showers temple, healthy Spa, vital Spa, the Sanotherm, turning it into a unique experience, uniting aesthetics to the highest and sophisticated wellness technology. It is an experience in which the architecture creates the adequate atmosphere to control the light as an element of serenity and the presence of mass volumes in a monochromatic space that predisposes the senses to tranquility.

NOMINEE FOR BEST DESIGN AWARD OF CLUB
最佳会所设计奖提名奖

胡俊峰&成志（中国上海）

获奖项目/Winning Project

拾院
With Nature

设计说明/ Design Illustration

天井的记忆

这里原本只是城市东郊一个垃圾场，我们用了3年植树造林，修复了这个地球小伤疤。建造了10间房，让人们可以在这儿停留、交谈。侥幸在这样一个环境，能让人提及些关于人与自然的话题，那将是我们最期待的事情。

人、自然、建筑，应该是怎样一种关系最恰当。记忆中，儿时外公家那个被四周屋围成的院子，是让我一直无法忘记的建筑形式。一群孩子围着天井追逐嬉戏，阳光、雨水、落叶、空气、鸟儿……透过院子上方的"开口"没有遮挡的进入空间。很想用这个案例来再现这段记忆。

与业主达成共识后，我们开始了。最终确定并完成的建筑平面，实在会让很多人提出各种常识性的质疑，下雨怎么办，需要走这么长的路去卫生间吗，厨房很远……在那段日子里，我反复权衡，功能一定是首要吗，习惯永远是必需吗……患得患失间，我放弃了人们口中的那些"绝对"，选择了内心对空间细腻和真实的体验。

为了使建筑"消失"，让它不再是人与自然的隔阂，我们用深加工处理后的土建余土，完成了建筑所有立面，而那些在旧木市场收购回来的老木头则是室内空间的主角。清晨，当第一缕阳光透过建筑的缝隙进入室内，洒在泥土和旧木的制造的墙上，光影让空间产生的变化和质感，让人兴奋不已。

几乎所有来过这个空间的人都能读出"中国空间"的感觉，这让我非常高兴，因为这的确是一个没有刻意却又很想传递的信息。我想，这可能是大家对"天井"都有一个共同的记忆吧。

Memory of A Patio

It used to be a trash dump in the eastern suburbs of the city. We spent three years planting forests to renovate this scar on the earth. In addition, we constructed 10 houses where people could have a short stay and talk with each other. If by any chance people mentioned a topic regarding human and nature in such an environment, it would be exactly what we are expecting to see.

What would be the most proper relationship among human, nature and architecture? Grandpa's courtyard surrounded by four houses in my childhood has been an unforgettable architecture style. A group of children chased each other around the patio where sunshine, rains, fallen leaves, air, birds accessed to space without any barrier.

With the passion to resurrect the memory, we started the project upon consensus with property owners. The final architecture floor plan aroused various doubts such as 'what if it rains heavily, is it necessary to walk that far to the bathroom, why the kitchen is further than before' etc. During that time, I considered back and forth and asked myself whether functions and habits should come first. In the end, I abandoned the so-called 'absolute' concepts, and listened to expectations and true experience from heart.

Aiming to 'disappear' architectures and any barrier between human and nature, we used processed residual soil to construct all elevations. Besides, those old woods purchased in the market are key elements in interior space. When the first sunshine penetrates through architecture into rooms in the morning, and shines on the wall made of clay and woods, people are excited to see the shadow that energized the space.

I feel truly pleased that almost all the people who came to the space could perceive the Chinese elements in it since that is exactly what I am trying to express unintentionally. Probably, there is a memory of a patio in each person's heart.

艾特奖

最佳样板房设计奖
BEST DESIGN AWARD OF SHOW FLAT

INTERNATIONAL SPACE DESIGN AWARD

117
IDEA-TOPS
艾特奖

获奖者/ The Winners
许盛鑫
（中国台湾）

获奖项目/Winning Project
反景入林/ House in the Green

IDEA-TOPS
艾特奖

获奖项目/winning Project
反景入林
House in the Green

设计说明/ Design Illustration

主动线的铺陈由碎石径至临水的架高木栈道，沿途错置的景石像是逗点一般，拉出与人行动线之间微妙的距离平衡，宛如形势天成的林间幽径，充满着迂回转折的味道。看似随意植下的数座建筑量体，沿着沁凉水池畔迤逦，藉穿梭于大面玻璃窗间的光影反映，争取最佳的通透、深邃之美。
刻意将木制桁架外露的尖顶建筑，有种引人合掌仰望的崇敬之意，选用香柏打造大门，上半部的桁架与两侧镶嵌玻璃，以及入内后侧向的建筑体取屋檐斜面的2/3到正立面，皆以大片玻璃镶嵌，顺势导入充沛天然光与树影。基地内彼此衔接的建筑体，利用明亮的透光步道串连，仰望镂空的斜顶木结构，释放仰角的高挑美感，大片面外的玻璃界面，以外突的白烤铁件强化支撑，但更精致同时彰显设计观点。建筑之一于屋顶中央凿开方形天井，四面悬空的双层木格栅凌空垂挂，仿佛双手托着自地面卵石堆里迎光生长的亭亭花树，在周边馨暖的木头香气里，演绎着比装置艺术更深一层的人文精神。

The wood pathway on the main route is fully covered with blinding, and the random rock in the landscape is like dots, connecting people. The view is breath-taking, you simply can't never get enough of it. From the refection, the buildings along the pond is so fascinating that the visit here becomes totally with worthy.

The building with pointy roof has some kind of esteem purpose. The front door is made of cedar and the frame is made of glass, and when you step inside the building, more than half of what you see is enormous embedded glasses, bringing in the nature sunlight. There's a bright pathway connecting these back to back buildings, so when people walk along, the pointy roof design, the sunlight coming in through the glasses, it's unlike anything you've ever seen in your life, absolutely stunning. And the design team made a square courtyard right in the middle of the rooftop, floating in the air, when looking up, it's like a delicate flower in your hand, fragile but full of hope. The design here well interprets the humanities that's beyond the installment art.

获奖评语

在朴素的传统房屋轮廓里，此项目娴熟地雕刻并镶嵌了许多种温馨的室内半室内空间。

This project successfully re-invented use of traditional materials like bamboo in making of contemporary space for relaxation.

NOMINEE FOR BEST DESIGN AWARD OF SHOW FLAT
最佳样板房设计奖提名奖

彩韵室内设计有限公司（中国台湾）

获奖项目/Winning Project

白色地景
Whitescape

设计说明/ Design Illustration

霁雨稀释了白色建筑量体的边界，轻盈、溢光——一座岛屿、一座山峦、一座雕塑感知的定律被翻、转、进、退，矩形样态的破格，宽朗的斜面屋檐是尺度的基线，围塑室内接区的视野景观，正立面右侧是矿石切面般的曲折流动，中央格栅罗列的角度与立面线条的走势一致，纹理如自然天生，虚掩的间隙透漏微光，在进入左侧门径之前，招引而出，本案作为地景建筑以几何意趣探勘空间涵构的意图。

移步室内，不规则几何外观过渡为一种均质的空间感，迎宾柜台后方是3间开放式接待区、1间会议室，廊道末端以接待柜台作收，另一侧是工作室、会议室，间隔着公开陈设的模型工学区，再者是盥洗室和机房；两间样品屋坐落于接待空间右侧。对应的开／隐比例，互为端景的接待区与模型区，柔和天光穿透其间；有序的公／私配置，渲灰立体石墙界分柜台与洽谈区，专用及隐密区域以木皮墙面收束在后，此外，接待柜台处以深浅色阶的木皮编织为墙面，接待区座位的隔屏，铁件格栅穿插如林木，配衬压低重心的花形灯具。光照是隐约澄净的，纷然意象在白色地景的内在，敞开一片森林。

The final showers have blurred the white architectural borders, creating a light and luminous halo of an island, a mountain, and a sculpture.

Established laws governing our senses have been turned around, transformed, evolved, and devolved, breaking down the conventional cuboid borders.

The broad slanted roof establishes the foundational geometric rule that surrounds and molds the view from within the indoor reception areas.

The surface directly opposite to the right displays the curves and flows of a freshly cut mineral stone.

The lines along the slant and elevated faces arrayed along the central corridor are melded into one contiguous entity with contours that could only be hewn through natural means.

Hidden gaps allow only the most subtle infiltration of light. Before entering the path on the left side,

Light would be drawn out to illuminate the area. Geometric principles and prospective spatial architectural concepts of the landscape constructs are fully lavished in this project.

As we move into the rooms, unregulated geometric shapes slowly give way to a homogenized spatial sensation. Beyond the reception counter are 3 open reception areas, 1 conference room, and another reception counter at the end of the corridor. A workshop and another conference room lie on the other side, separated by an exhibit with publicly displayed engineering models. This is then followed by the bathroom and machine room. The two model homes are located on the right side of the reception area. The ratio of open / closed areas, reception and model exhibit zones that act as complementary features with gentle natural lighting showering the spaces have been perfected. There is an orderly arrangement of public and private spaces, while shaded ash walls of stone separate the counter from the discussion areas. Specialized and private areas are tucked away with walls lined with interwoven wooden bark. Partitions and iron railings separating the reception seats are arranged like serene forests complemented by flowery shaped lights brought closer to the negotiators to provide only the purest illumination, opening the forest before us to unveil the whitescape of our imaginations.

NOMINEE FOR BEST DESIGN AWARD OF SHOW FLAT
最佳样板房设计奖提名奖

彩韵室内设计有限公司（中国台湾）

获奖项目/Winning Project

弥合之境
The Realm of Confluence

设计说明/ Design Illustration

疾行的节奏，至此都和缓，四方萦绕的池水、风雅的古旧陶瓮，点映于实木与天然石共构的建筑主体，量体大而蓄涵，接引方直而生动的内外线条，材料应着空间展露其裸生的不伪然质地如诗词之衬字，扬抑语气，眺凝意境；方寸庭院里的疏朗木枝，则彷佛凭助了量体的立直端稳，自回廊进门之际，倾听便有近处城市烟尘冉冉退隐于时空，那微细音响。

木、石等自然元素，透过大面积落地窗毫无障碍的引入室内，空间的配置朝横轴拉展，自有舒张之意气；廊道两侧，接待柜台与洽谈区域隔着黑石勾边的木质地坪板材互应排铺，主空间右侧，可经由户外沿廊曲径通往视听放映室与模型展示空间，左侧则邻接三间样品屋。整体室内空间的布局及动线依使用机能作出明确分划：厅堂、回廊、院落、边间，再现古典大家宅邸的行走经验；置中、转缓、明快的空间格局，取自现代设计的收弛有度。接待柜台延续大厅廊地道坪以异素材拼接的立体设计，石与木拼接的纵长块体，浅深色度、异质纹理，宛若山林背景之中，云雾流泻；其实际效用则在于隐晦且优雅的暗示出主控柜台区位。四座石板墙作为个别洽谈区的隔分屏幕，精心量制的厚度恰可嵌入平板屏幕，深色木桌、淡灰沙发绝无奢华高调，而是在望向窗边由交织铁件所框画的水景与碧景，彼时，随同心境，自身、外物亦不再截然有别，如同这座侧居高楼群落一隅的水榭亭阁，悠然隐于市。

Every bustling and rushed schedules will be brought to a gentle pace in this realm.

Surrounded by serene pools and streams on four sides, the architectural features are decorated by elegant antique ceramic urns, trees, and natural stone.

It is massive and connotative, connecting with straight and vivid internal and external lines. Responding to the space, material displays is natural and true texture.

The word complementary has been fully realized in this realm of visual poetry.

The courtyard and its trees are given a sense of gravity with the geometric edges of the central architectural construct.

Upon entering the corridor, the background urban noises and smells give way to faint murmurs as we enter another time and space.

Wood, stone and other natural elements are brought indoors through large French windows. Spatial arrangements extend horizontally, giving visitors a sense of relaxing width. Both sides of the connecting corridors, the reception counter and reception areas are compartmentalized with inter-crossed wooden flooring and panels lined with dark stone. The right side of the main exhibit can be accessed through the outdoors corridor that also leads to the AV display room and model exhibit gallery. Three model houses are linked at the left side. The entire spatial arrangement and movement lines of the indoor space have been distinctively compartmentalized, with the main hall, corridors and side rooms recreating the walking experience found only in classical manors. The moderate, gentle, and dynamic spatial elements have been inspired by the flexibility of modern design features.

The reception counter is a three-dimensional extension of the flooring of the main hall corridors, composed of a different yet perfectly complementary material. An extended mass of arrayed stone and wood facings with contrasted tones and textures paints a canvas of billowing clouds across mountains and forests. The design objective is to provide a subtle and elegant focus to the main reception area. Four stone-faced walls with carefully optimized thicknesses are used to partition the discussion areas. Dark wooden tables and pale ash sofas provide a most luxurious experience. One can look outside the window to admire the interwoven iron railings outside the walls and the emerald waterscape it borders and realize that the individual and the surroundings are no longer separate, but are melded as one as we become part of this terraced pavilion hidden away in a corner of the concrete jungle.

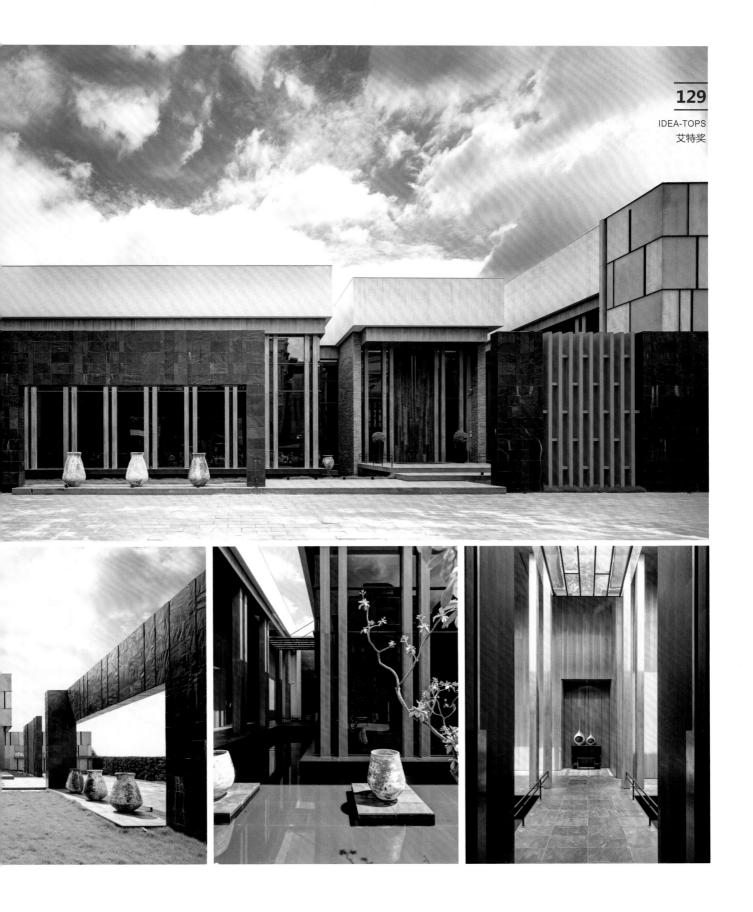

131
IDEA-TOPS
艾特奖

NOMINEE FOR BEST DESIGN AWARD OF SHOW FLAT
最佳样板房设计奖提名奖

邵唯晏（中国台湾）

获奖项目/Winning Project

棋琴文鼎苑
Symphony of Chess and Zither

设计说明/ Design Illustration

古典美学再转译

本案位于南台湾高雄某建设公司的住宅公共空间，经过多方的讨论，决议本案有两个写实的前提，第一是业主方希望设计上有古典的元素出现。第二是业主期望本案机能上除是一个住宅的公共空间外，希望导入博物馆的氛围。故在设计主轴上将创造一个生活与展示的交融场所，并尝试在古典美学的基础上重新再诠释，放掉古典既有的包袱，用较写意的方式，将古典建筑抽象化、元素化或符号化，并透过当代的手法重新转译，巧妙地融入空间中，希望重新联结现代简洁与古典雅致的对话，简洁有力的当代线条和语法中带有朴实高雅的古典优雅。

流动的空间格局

本案的公共空间很宽广，在平面布局上和对于建筑结构的处理态度上，我们有几个开放式的做法。在平面配置上，基本上分为两大区块：前厅和后厅，我们企图打造一个流动的空间，故将隔间墙体全部拿掉，空间中找不到任何的隔间，并导入一条展示的动线，希望将建筑的前后打开重新串连，并创造室内外关联的最大可能。在后厅的部分也将建筑外墙拿掉，让最大量的光线与风能够流窜进来，所以后厅基本上是一个半户外的过渡空间，并且将每一种机能（如健身房、儿童游戏间、交谊室等）放在一个个的盒子里，悬挑的结构配合透过轴线的翻转和盒子的错迭，企图创造出一个无落柱的半户外开放场域。

古典柱式再诠释

古典的元素优雅而细致，我们导入当代美学的手法——形变，模糊并融接古典样式中的顶、身、座等明确元素，透过形变的手法重新诠释古典柱式。同时在几个重点端景上，语汇上只保留其重要的弧形和曲线，透过几条简单线条重新勾勒，将三维的变化转成二维的纹理，刻意营造出既古典又当代之空间经验，创造出一种极富张力的和谐感。对于建筑外廊的天花也是相同的手法，将壁柱与穹顶透过形变的手法，重新勾勒保留其断面线条，模糊融接的当代语汇中带有古典雅致之感。

古典的数字艺术转译

本案的艺术品都是特地为响应本案而量身订做的，在设计上遵循空间设计的脉络，将古典的元素透过数字的手法转译。"风"系列作品希望捕捉自然能量的有机生命力，并运用当代参数运算模型，结合3D打印的新创作方法，以连续流动的几何诠释古典语汇的新生命与活力，创作过程中以辐射状同心圆之点线面数学分割关系，不断进行曲面实虚形变的数字拟态，层层蜕变为理性与感性交融的独特有机容器。尝试融合古典元素之质感气韵，与当代数位衍生几何的美学秩序，体现古典之新义。

Classical Aesthetics Retranslation

The case is a public space of a construction corporation which located in Kaohsiung, southern Taiwan. After many discussion, there are two realistic premise of the case, first is the appearance of classical elements as the owner expected. The second is importing the atmosphere of museum to make the public areamore than abuilding for living.Thereforeour principal point of design is to create a blended ambience of life and demonstration.We attempted to re-interpret the space based on classical aesthetics. In order to achieve the goal, wereplace the burden of classical patternwith more impressionistic approach to abstracting, element converting, symbolic it. Through the contemporary approach we re-translatedthe space with these ingredient, hoping to reconnect the dialogue between modern simplicity and the classical elegance. The design accomplished the humble and elegant classical style concealed inuncomplicated contemporary lines.

The spatial pattern of flow

Due to the public space is verywide and broad,we had a few approach for the concept ofmanaging configuration plan and architecture structure. For the configuration, basically it divided into two area: Ante hall and Back Hall. We were seeking a method to create a flowing space, thus we removed all of the compartment walls, afterward the compartments is no longer been found in it. Moreover we insert anexhibiting pathin this zone, opening the space of architecture in addition relinked the front and back yard, bring aboutthe maximum possibility of producing the connection between indoor and outdoor spaces. In Back Hall, we tear down the facades, allowing the maximum amount of light and wind flows can spread in.Thence the Back hall is basically a semi-outdoor transitional space. We attempt to create a semi-outdoor opened field with no column by settingevery function area (such as gym, playground for children, recreation room, etc.) in each box and combining the cantilevered structure, the rotation of axis and boxes staggered to attain the goal.

Reinterpretation of classical column

The element of Classical Aesthetics is elegant and delicate. We insert the technique of contemporary aesthetics – Morph, into blurred and merged classical pattern for the top, body, seat and other explicit elements toreinterpret classical column. On several light spot, we only left the significant arc and curve of it.By drawing a few pure lines, it turned a three-dimensional intotwo-dimensional pattern. We made an effort on creating a space experience of classicalcoexist with modernas well as a sense of

harmony and tension. For the ceiling of corridor outside the building, we duplicated the same method. Through the skill of Morph, the pilasters and dome retain its cross-section lines but re-sketched. Makingtheblurred and merged modern pattern concealed the sense of classical elegance.

Classical digital art translation
The artwork for this case are tailored specifically.Following the spirit of interior design, we tried to convert the classical elements through digital skills. The series works called "Wind" appeal for capturing the organic vitality of natural energy. We used the modern computing parameters combined with new creative method of 3D printer, bringing out a new life and vigor for the classical pattern by continuous flow of the geometrical interpretation.During the creating process, we took advantage ofthe point, line, and surface concepts of mathematical from concentric circle tomake shape dividing and reforming.After repeatedly digital simulationsof curve revealed and concealed, the organic container turned out to be a unique combination of rational and sensibility.We trying to merge the qualities, characteristic of classical elements and the order of geometric aesthetics derived from contemporary digital, reflecting the Classicisma new implication.

NOMINEE FOR BEST DESIGN AWARD OF SHOW FLAT
最佳样板房设计奖提名奖

唐列平（中国佛山）

获奖项目/Winning Project

天湖郦都《裹中》
Likeview Garden Guozhong

设计说明/ Design Illustration

从空间中去寻找平衡，人与空间的关系、与大自然的关系，设计者希望透过这种心境，营造一种比较内敛的生活方式，通过轴线关系，使用者无论走到空间的任何角落，都可以感受到空间带给我们启发，感受是最直接的，但并不张扬。设计同时围绕"合"的概念进行叙述，每个空间都应具备节奏感，虚实有度，引导居住者衍生出一种真实的生活态度，围绕建筑核心本质而营造空间。

In seeking space balance and relationship between people, space and nature, the designers hope to develop a restraining lifestyle via the axial line relation. Users can sense the enlightenment brought by the space at any of its corners. Such a sense is direct and low-key. The design also centers on the concept of "integration" for narration. Each space should be rhythmic and well tailored, making residents have a true attitude towards life. The space is built by centering on the core building essence.

140
Best Design Award Of Apartment
最佳公寓设计奖

艾特奖
最佳公寓设计奖
BEST DESIGN AWARD OF APARTMENT

IDEA TOPS
INTERNATIONAL SPACE DESIGN AWARD

141
IDEA-TOPS
艾特奖

获奖者/ The Winners
俞佳宏
（中国台湾）

获奖项目/Winning Project
墨方/ Square Ink

142

IDEA-TOPS
艾特奖

获奖项目/Winning Project

墨方
Square Ink

设计说明/ Design Illustration

轻抚山川流水刻划出的纹路，
寄语生活人文的原色，
光温润的晕染上原木质地，
细腻雅致映入凝视的目光。

为一位拥有诗书气息的单身女性，细腻地描绘如书画笔墨一般，蕴含人文气息的居住空间，格局配置上，强调简洁方正；颜色氛围上，则着重于静谧与雅致；而整体的视觉观感，更是倾心于空间连贯的对称性。

以铁件的沉稳质感，将一块拥有大自然色泽与肌理的石材，围硕成一个完美比例的长方框型，壁炉上的大理石，流水般的线条将公共空间的利落刻划出间接性的区隔；而拥有整个空间的核心价值，与温暖的内含光的温度，名为"泼墨山水"的大理石，将之置于客厅与书房、客厅与餐厅的中轴线，沉稳地带出原木温暖的色泽，与拥有大地触感的空心砖墙面。

Touch the fine lines carved by mountains and rivers
Expression of the original color of life and humanity
Log texture coupled with mild lighting
Fineness and elegance full of gazing eyes
Like a scholarly single girl who is delicately depicted as painting and calligraphy ink, a living space with deep-rooted humanism factors should emphasize conciseness and neatness in pattern configuration, lay emphasis on tranquility and elegance in color atmosphere, and focus on the symmetry of spatial coherence in overall visual perception.

Based on the calm texture of iron accessories, a stone with natural color and texture is enclosed into a rectangle with perfect proportion. The flowing lines of the marble on the fireplace carve out indirectpartitions for the agility of the public space. The core value of the whole space lies in the middle axle between living room and study and between living room and dining room with warm light temperature and "splashed-ink landscape" marble, bringing the calmness to the warm color of log and the air brick wall with a sense of earth.

IDEA-TOPS
艾特奖

获奖评语

现代的洗练空间，一致而又层次分明的细节。

Modern and concise space, consistent and clearly structured details.

NOMINEE FOR BEST DESIGN AWARD OF FLAT
最佳公寓设计奖提名奖

Lidija Dragisic（斯洛文尼亚）

获奖项目/Winning Project

几何寓所
Geometric Residence

设计说明/ Design Illustration

该公寓坐落在斯洛文尼亚的首都卢布尔雅那的中心区。它分为居住区（包含一个大型客厅与厨房和饭厅）以及更亲密的睡眠区（两个睡觉的房间和一个书房）。在这两个空间之间，有一个服务部门提供两个现代浴室和一个公用设施。

当客户找到我们时，这个公寓已经搬迁并已损毁了。为了把它带回到现实生活中，改造是绝对必要的。这种改造作业并没有对平面图做出任何重大的改变，除了一些浴室的改进，窗户和门的更换，新铺设的地板和电气安装。

室内装饰设计很简单，设计方案结合了基本材料和天然颜色：橡木、白色家具以及在一些细节上注重黑色。这种方法允许居民自由进行任何额外的装饰（艺术、彩色靠垫、地毯、饰品等）而不会影响整体设计的特点。

沿着墙壁的家具采用一种中性的白色，并采用一种几何图案进行设计，这样放大了该公寓的氛围。几个精心挑选的重点是木材，这能营造一个愉快和舒适的气氛。家具是定制的和独特的，例如，一个大型3.5m长的木桌位于餐厅区域的中间或满足宴会需求的一个巨大的厨房，这将生活空间和私人空间汇集在了一起。在主卧室实行白色和橡木的联合（主卧室和一个几何形状的墙衣柜在上面）。在整个公寓的巨大窗帘实现了额外的亲密感、柔软性和一致性（它们也是采用一种中性的黑色和白色组合）。浴室很简单，没有任何不必要的装饰，但具有美丽的混凝土一样的瓷砖，因而与众不同。通过巧妙利用内置的镜面墙，这些空间在视觉上得以增强。

我们的目标是使得室内适应多方面的、现代客户不断变化的需求。除了功能性以外，要创建许多不同的照明场景和环境。这是通过结合基本照明与间接LED照明而实现的，这些隐藏在家具元素里面。在卧室里的天花板上装饰着水晶吊灯，因此（连同其他古董配件）赋予了这个住宅一个独特的灵魂。

The apartment is located in the heart of Ljubljana, the capital of Slovenia. It is divided into living area (consisting of a large living room with kitchen and dining room) and more intimate / sleeping area (two sleeping rooms and a study room). In between these two spaces there is a service part offering two contemporary bathrooms and one utility.

When the client approached us, this apartment was vacant and ruined. In order to bring it back to life, the renovation was absolutely necessary. The adaptation didn't make any significant changes to the floor plan, besides some bathroom enhancements, windows & door replacements, new flooring and electrical installations.

The interior furnishing design is simple, with design-scheme combining basic materials and natural colors: oak, white furniture and some detail accents in black. This approach allowed the residents the freedom of doing any additional decorations (art, colored cushions, carpets, accessories etc.) without disrupting the overall design-statement.

The furniture along the walls is in a neutral white and designed in a geometrical pattern, which magnifies the flat's ambiance. Several carefully selected accents are wood, which create a pleasant and cozy atmosphere. The furniture is custom and unique – for example, a big 3.5 m long wooden table in the middle of the dining area or a huge pantry with parquet-finish, which brings the living and private spaces together. The marriage of white and oak is implemented in the master bedroom as well (master-bed and a geometrically-shaped wall closet above). The massive curtains throughout the apartment achieve additional intimacy, softness and consistency (they are also in a neutral black & white combination). The bathrooms are simple and without any unnecessary decorations, which allows the beautiful concrete-like tiles to stand out. These spaces are visually enhanced with the clever use of built-in mirrored walls.

Our aim was to adapt this interior to the versatile, ever changing needs of the modern customer. Besides the functionality, we wanted to create many different lighting scenarios and environments. This is achieved by combining basic lighting with the indirect led lighting, which are hidden in the furniture elements. The ceiling in the bedroom adorned with a crystal chandelier, which (along with the other vintage accessories) gives this residence a unique soul.

NOMINEE FOR BEST DESIGN AWARD OF FLAT
最佳公寓设计奖提名奖

黄翊峰（中国台湾）

获奖项目/Winning Project

人文寓景
Humanity Space

设计说明/ Design Illustration

清新寓景人文感度

北欧风格已是当代设计的重要潮流之一，如何赋予更为独特的个人品味或是丰厚的人文素养，是许多设计师的挑战，亦为本案意欲具化成型的设计构思；对于爱好旅行的屋主来说，通行全球的北欧风尚，如某些精致的设计旅店一般亲和、隽永。

甫进入屋内，清爽的浅色板岩砖地坪、木制家具展布开来，以多机能的隔层／橱柜块体，简洁分划空间属性且兼顾收纳功能，客厅沙发区背墙与红酒柜围构出玄关，墙体除了可作衣帽柜使用，表面架设烤白的格状铁件搁板，加入展示用途，再搭衬以工业风灯具、设计感家饰，精准展现北欧设计巧思的内敛表现形态。电视主墙则作为公共场域用以聚焦的视觉亮点：在清水模砖面之上，蓝色与灰褐色的木纹砖拼组为墙景，低抑的彩度，使设计不过分影响生活实景；清水模砖面的低调质感，则可无碍融接隐于铁件勾边玻璃门后的书房空间，设计师在此有意识使用通透材质，装饰性降至最低的书房，成为接引自然光的介质，连贯风格之余，仿佛拓延了公共空间纵轴的深度，亦解决了长廊的采光问题。

隐私空间的布局撷取了木屋意象，倾斜天花让重心缓缓下降，并隐去了大梁的存在，强调卧室的沉静感受；床尾板材拼接，呼应电视主墙的设计手法，且以单色取代缤纷，反复烘托舒适柔和的气氛。自外而内，风格与生活相互容纳，创造的已非某种设计趋势，而是通连个人生命真实情貌的日常。

Fresh sense of Humanity

Nordic style is one of the important trends of contemporary design. And it is a challenge for many designers to give more unique, personal taste or rich cultural literacy, also is the intended idea for this case. For the house owner who loves travel, the prevailing global Nordic fashion, namely, is as meaningful as some sophisticated design hotels.

Entering the house, the room spread out with the floor of fresh and light-colored slate tile and wooden furniture, it is divided into different areas by multi-function compartments and blocks cabinets which already taking into account the needs of storage function. The living room entrance is structured with the back wall near sofa and the wine cabinet. The wall is used as lockers and display area, with the iron piece of white trellis shelf, loft style lighting and design furnishings, precision showing Nordic design ingenuity manifestation. The visual highlight of public area focus on the main TV wall. Blue and taupe color of wood-finish brick on the Architecture concrete wall, with low saturation, which does not affect the real lives, understated texture of Architecture concrete connect smoothly to the study behind the glass door.This designers use transparent materials consciously to minimize the decoration of the study, and become the media of natural light. It not only stretches the longitudinal axis of the depth of the public area, accordance with the design style, and also solve the lighting issue for long walkway area.

The layout for private space is imagery of chalet. The gravity down slowly by the tilted ceiling, and withheld the existence of girders, emphasizing the quiet of bedroom; the spliced bed plate echoing with the design techniques of main TV wall; and repeated in contrast with soft and comfortable atmosphere and replace the monochrome colorful by monochrome. From outside to inside, life style and life accommodate to each others. It creates not just a design trend, but true appearance of personal daily life.

NOMINEE FOR BEST DESIGN AWARD OF FLAT
最佳公寓设计奖提名奖

张祥镐（中国台湾）

获奖项目/Winning Project

適性居所
Causal Life

设计说明/ Design Illustration

沿着橡木、轻巧云彩、朴实岩质，
漫步在开放通透的公共领域空间，
轻诉着与自然间的对话，
搭载量身完善的生活机能，
恬淡适己身心自在。

Along the oak wood, light cloud pattern, simple stone material
Walking though open filed space
It seems can dialogue with nature
Within customized perfect living function
Simple life , casual way

157
IDEA-TOPS
艾特奖

158
IDEA-TOPS
艾特奖

NOMINEE FOR BEST DESIGN AWARD OF FLAT
最佳公寓设计奖提名奖

罗仕哲（中国台湾）

获奖项目/Winning Project

延伸出開放空間的深邃感
Profound

设计说明/ Design Illustration

和谐的深浅用色拿捏
此户型的开窗多且面积大，运用深色可带出空间的独特性。胡桃木地板奠定沈稳、舒适的氛围。客餐厅之间的横梁选用近于墨黑的深蓝色。欣盘石在3D模拟的过程发现此处若刷成深蓝，视觉效果会比惯见的白色更佳。一条条的硅酸钙板天花为优雅的浅灰蓝，周遭勾勒一圈深蓝；再搭配浅岩色调的立面，整体用色在轻重、冷暖之间皆获得平衡。

藉由线条来营造深邃感
半穿透的造型天花让客餐厅显得高敞。原始天花漆成深蓝，下方则以一条条硅酸钙板构成镂空的造型天花。原始天花从硅酸钙板之间露出、形成超长的深蓝线条；视线顺着这些长线条而延伸，无形中拉大了空间深度、强化此区的宽敞感受。此外，硅酸钙板的间距宽窄不等，让简练、理性的线性造型也能活泼地展现变化。

远观近看各有风貌的石墙
设计师很讲究居住者的心理感受，总是避免选用那些会引发视觉疲劳的色彩或质感。客厅单侧的大理石墙舍弃带有结晶的石材，以免炫目反光形成不适。这道墙，远看以为是素面，近观则会发现它夹杂了化石跟贝壳。半凿面的处理佐以几条装饰线，当洗墙光投射而出，细微变化的光影构成了迷人背景，也满足屋主崇尚自然与温馨的品味。

活用拉门来变化感受的柜墙
这对夫妻并没有看电视节目的习惯，为了来访亲友，他们仍在客厅配置电视主墙，只希望平日能遮住这台电视。欣盘石沿长墙打造简约的造型柜墙，内为大量的储物柜与电视墙。藏有电视的吊柜选用可以完全收入的巴士拉门，不论观看或遮蔽电视，立面都能保持平整。客厅另一端的书房柜墙，则运用可自由横移的大拉门来变化空间感。

随时感受家人存在的私角落
男主人需要一处能不受干扰的书房，以便阅读或上网。但他又希望能了解家人在其他空间的动态。设计师将餐厅后方的卧房改成书房，以铁件、大理石与灰玻璃构成了"⊓"字形屏风。适度的遮挡让男主人享有独立角落，还能随时感受家人在隔邻的动态。男主人在入住后对欣盘石表示：他非常满意这样的贴心设计。

便于料理与出餐的餐厨空间
女主人习惯在下锅前将食材全处理好并分装在容器，再从容地加热、出餐。她因此想要有座中岛，以便能有充裕台面来备餐。欣盘石在厨房的中央设计配置大型中岛，满足了食材从冰箱移至炉灶的过程中的各项需求。这座中岛与餐桌位于同一条轴线，便于出餐。厨房入口并装设自动感应的电动门，无论出餐或收餐，出入都更便利、安全。

Harmonious use of dark and light colors
The windows of the apartment are designed in a large number and area, with the application of dark colors to show the uniqueness of the space. The walnut floor helps create a calm and cozy atmosphere. Deep blue color close to black is used for the beams between the living room and the dining room. In the process of 3-D simulation, CS Design discovers that if it is painted with dark blue color, the visual effect will be better than white. The calcium silicate ceiling draws a circle of dark blue color around the elegant light grayish blue color; matched with the façade of shallow rock tone, the overall coloring gains a balance between dark and light and between cold and warm.

A sense of depth created by lines
The translucent modeling ceiling makes the living room high and spacious. Painted with dark blue color, the original ceiling adopts calcium silicate boards on the bottom to form a hollow modeling ceiling. The original ceiling is exposed from calcium silicate boards and form overlong dark blue lines; with the extension of sight along these lines, the depth of the space is increased and a sense of spaciousness is intensified imperceptibly. Moreover, the distances between calcium silicate boards are different, which makes the concise and reasonable linear modeling show variation vividly.

A stone wall with diverse landscapes
CS Design concerns a lot on the metal feelings of the dweller and always avoids using the colors or textures causing visual fatigue. Stones with crystal are abandoned for the marble wall at one side of the living room lest discomfort from reflection. This wall, when seen from afar, is like a plain surface; when seen closely, it is mingled with fossils and shells. The half cut surface is coupled with several ornamental threads. When wash wall lights project out, the subtly varying shadows constitute an enchanting background, which caters for the taste of nature and warmth of the owner.

Cabinet walls equipped with sliding doors
This couple does not have the habit of watching TV. However, they still require setting a TV wall in the living room, merely hoping to cover up TV set. CS Design sets a concise modeling cabinet wall along the wall for TV set and other objects. The wall cupboard for holding TV set adopts a bus sliding door to keep the façade neat. The cabinet wall of the

study at another end of the living room uses a large sliding door to highlight spatial variations.

A private corner accessible to know the activities of other family members

The host requires an undisturbed study for reading or surfing, yet he hopes to get to know the activities of his member members in other spaces. The designers transform the bedroom at the back of the dining room into a study and create a ⊓-shaped screen with iron accessories, marbles and grey glass. Proper shelter enables the host to enjoy an independent corner and get to the activities of his member members. After occupancy, the host expresses to CS Design that: he is very satisfied with such considerate design.

A kitchen space convenient for food preparation and delivery

The hostess usually processes food materials and puts them in different containers before have them heated and delivered. Therefore, she requires setting a central island for food preparation. CS Design sets a large island at the center of the kitchen, which meets various food preparation requirements from refrigerator to cooking range. This central island has the same axis with the dining table, which is convenient for delivery. An electrically operated gate is set at the entrance of the kitchen, which makes delivery and collection more convenience and safe.

艾特奖
最佳办公空间设计奖
BEST DESIGN AWARD OF OFFIC SPACE

INTERNATIONAL SPACE DESIGN AWARD

165

IDEA-TOPS
艾特奖

获奖者/ The Winners
Moura Martins Architects（葡萄牙）

获奖项目/Winning Project
Famo办公大楼 / Famo Office Building

获奖项目/Winning Project
Famo 办公大楼
Famo Office Building

设计说明/ Design Illustration

本设计将位于葡萄牙北部洛扎达的Famo办公大楼进行了全面整修。

该三层建筑需要进行内部重建，以便界定各个区域，改善各部门之间的交互作用，并分清各个路径和通道。

在每一层都重新建设了一个大厅。在中间的楼层上，这个大厅与生产设施进行视觉交流。这个走廊在每一端都有一个楼梯，这便于各楼层之间的通信联络。

在该中心，来到该大厦内部的另一个垂直通道，这标志着该建筑物的立面。这些楼梯，从该建筑的原始设计，也重新进行了整修，使得曾经失去的通道又重生了。

顶层的通道是用于管理办公室、财务会计部、培训室和酒吧。两个空间被一个可移动的墙壁分开，从而扩展了这些空间的功能。

行政办公室沿着主要走廊进行部署，被玻璃分区和橱柜分开，这样保护了必要的隐私。

在中间的楼层以上，是接待区域和操作区域。建筑的立面通过楼梯加以标记，以便让我们找到接待区域所处的位置。

这个区域，对附近的工厂具有独特的视角，因而使每个人感到惊喜。

另外，在这一层上，整个外壳屋顶建筑被占据，出现了有关咨询、销售和设计团队的工作区域。这个开放空间在该中心具有3个封闭的模块，可以容纳普通空间，比如会议室、打印机房间和技术设施。

在该中心创建了一个空白区域，其具有一个独立的楼梯，从而使得这个空间能够与位于一楼的展厅进行交流。

作为一家生产办公家具的公司，Famo工作区也是展览空间的延伸。该空白区域开放空间和一楼展厅建立起了一个直接的联系，因而改善和优化了这两个区域。

It was held a full refurbishment of Famo office building in Lousada, north of Portugal.

The three floors building needed an interior reorganization in order to define areas, improve interaction between departmentsand clarify the paths and access.

It was created a main hall that reappears on every floor. On the middle floor, this hall communicates visually withmanufacturing facilities. This corridor has a stairway at each end facilitating communication between floors.

At the center, it comes another vertical access within the tower that marks the building's facade. These stairs, from thebuilding's original design, were also remodeled, being reborn an access that was lost.

The top floor is aimed at management offices, financial accounting department, training room and bar. These last twospaces are divided by a removable wall allowing to extend the space for either function.

The administration offices are deployed along the main corridor, divided by glass partitions and cabinets that bring thedesired privacy.

In the middle floor arises the reception and operative area. In addition to the tower, the facade of the building is markedby a staircase that leads us to the place where the reception is.

This area is an ante-camera to the main hall that surprises anyone with peculiar views of the adjacent factory.

Furthermore, on this floor, occupying the entire shell roof building, appears the working areas of consulting, sales anddesign teams. This open space has at the center three closed modules that accommodate common spaces like conferenceroom, printer room and technical facilities.

At the center was created a void with a self-supporting stair sculpture, allowing this space to communicate with theshowroom located at ground floor.

As a company that produces office furniture, Famo workspaces are also an extension of the exhibition space. The voidallowed the creation of a direct link between the open space and the ground floor showroom, improving and streamliningthe two areas.

获奖评语

一个极其优雅而知性的氛围,不寻常但确定有效地刻画了这些办公室的特点。

An extremely elegant and spiritual atmosphere, unusual-but surely effective characterizes the intentions of these offices.

NOMINEE FOR BEST DESIGN AWARD OF OFFIC SPACE
最佳办公空间设计奖提名奖

Golden Ratio Collective Architecture（希腊）

获奖项目/Winning Project

金律总部
Golden Ratio Headquarter

设计说明/ Design Illustration

Golden Ratio建筑公司的办公室展现了实现沟通与合作的另一种方式。门口区域，大面积的木质表面营造出热情的氛围，欢迎着人们由此进入公司的内部空间。入口处有三个安全帽和许多公司成员的头像，由金属螺丝制成的令人印象深刻的logo指引着人们进入工作区域。在墙面上，你会看到"Two quantities are in the golden ratio if ..."的字样，紧接着便是Golden Ratio的办公空间。走上楼梯之后就到了特别设计的"壳体"，该空间可供不同部门和部门之间开展各种各样的活动和协作。墙壁和倾斜的屋顶运用了统一的木质表面，形成一个封闭的核心，保护了客户与公司的隐私，使人自信且营造了亲密的氛围。两侧及屋顶上的开口将自然光线引入室内，体现了Golden Ratio公司富于变化的特点和灵感来源。在该房间的一侧，光线透过金属结构的书架照射进来，通过网格的变化形成了颠覆性的设计。长方形的铂金灰工作台可用于会议，凸显着空间的几何结构；储物柜采用了相似的色调，使得整个区域具有完整性；简单而富有动感的线条体现出设计的沉稳和真诚，就像该公司对待客户的态度一样。

IDEA-TOPS
艾特奖

The Golden Ratio architectural office is an alternative approach to the concept of communication and cooperation. A broad wooden surface forms a welcoming gate to the inner space of the company. Three construction helmets, as many as the company members, along with the imposing logo made of metal screws, mark the entrance to the working space. "Two quantities are in the golden ratio if ..." and journey into the world of Golden Ratio begins. Following an upward course, the visitor reaches the specially designed shell, open to all kinds of activities and to collaboration between different disciplines. A single wooden surface covers the walls and the sloping roof forming a closed core, which shields the customer relationship with the company by creating a climate of confidence and intimacy. The openings on both sides, as well as those on the roof, allow natural light to invade space, expressing the evolution and inspiration, characteristics of the Golden Ratio. On one side of the room light is filtered by the metal construction of the library, which through the change of grid results in subversive design. The elongated anthracite metal workspace hosts the meetings, while at the same time emphasizes the geometry of space. Lacquer cabinets of similar colour discreetly complement the composition. The result is characterized by simple and dynamic lines, expressing a design stable and sincere, like the relationship of the company with customers.

IDEA-TOPS
艾特奖

NOMINEE FOR BEST DESIGN AWARD OF OFFIC SPACE
最佳办公空间设计奖提名奖

深圳市矩阵室内装饰设计有限公司（中国深圳）

获奖项目/Winning Project

重庆万科办公室
Chongqing Vanke Office

设计说明/ Design Illustration

每个大的集团公司都有自己的标准和个性，万科也不例外，在完成好基本的功能需求外，设计师尽量把万科精神注入到空间设计中去，传承、舒适、简洁，"以人为本"，把人作为设计的重要参考物，重视环境人性化解决之道，强调空间合理布局及细节的关怀设计。把以往对于办公空间的刻板印象进行一些升华。

Every large group company has its own standard and personality, so does Vanke. After meeting the basic function needs, the designers try to inject the spirit of Vanke into space design for an "inheritable, comfortable and concise" effect. The designers also follow the principle of "people oriented", focus on the humanization of environment, and emphasize reasonable space layout and considerate detail design, upgrading the stereotype on office space.

NOMINEE FOR BEST DESIGN AWARD OF OFFIC SPACE
最佳办公空间设计奖提名奖

Mocoloccomocolocco（波兰）

获奖项目/Winning Project

Onet室内设计
Interior office designe Onet.pl

设计说明/ Design Illustration

波兰门户网站 onet. pl渴望其排名在谷歌之上不是什么秘密。"谷歌办公室"几乎在世界各地都已得到认同，其具有非正式的空间是众所周知的。投资者希望其办公室内部也不寻常、充满活力、并保留在内存中，这与门户网站是有关联的。

设计者的任务就是基于非正式的途径进行设计和联络，从而大大增加员工们的创造力，同时在他们专心致志工作的空间里，给予他们放松的时刻。

这个任务是相当困难的，因为波兰门户网站onet. pl的办公室在一个4500m²的作业面上雇佣了650名员工。为了在功能上安排空间用于谈话、会议和工作，提出了各种想法和指导思想，其中包含遍布于一栋办公大楼的7个楼层的所有方面，与此同时，品牌被认为是该门户网站。

"灵感"成为我们周围的元素，与此同时，这些也是该公司的各个部门所具备的特征。独特性、轻盈性成为我们自己空间的特色，主要通过使用非装饰性材料、家具（主要是波兰设计者设计），使所设计的空间总是具有功能性且符合人体工程学。

一楼混合了地板颜色黑色和黄色标志而作为白色的背景颜色。接下来，参照这些元素的颜色。所以，我们有火、水、土地和空气。对于每一种地板材料，其结构和颜色都适合于该元素。

在大多数楼层，天花板是开放的，它们与设备都被刷上该元素的颜色。实际上，在大多数年轻人工作的空间里，这样的解决方案有助于获得更多的空间和阁楼办公室。

在一幢办公大楼的第一层楼上，一种常见的空间是用于员工娱乐室。在这样一个地方，你可以享受一杯好咖啡，坐在一把舒适的椅子上，拍摄照片，甚至可以玩弹球游戏机，而与此同时，在这个场地里，你还可以安排一个会议。

员工履行其职责，必须坐在办公桌旁吗?清除掉所有环境干扰或放弃休息，员工的工作效率就会提高吗?我不这么认为。经验表明，事实恰恰相反：你待在电脑前的时间越长（通常在这种地方只是有太多的……），关注力就越差；坐在椅子里的时间越长，就越疲倦；每天工作的情形越类似，工作的热情和履行职责的渴望越低。

一个不寻常的室内环境放松了僵硬的思维模式，并激励员工为客户着想。

It is no secret that the portal onet. pl aspires to rank on google. Google Office recognizable arepractically all over the world, are famous for their informal space. Investor wanted his officeinteriors were also, unusual, energizing, remaining in memory , associated with the portal.

The task of the designer is our task was to design and connect over the informal, which haveboosted the creativity of employees and give them moments of relaxation with spaces to work in theintently.

This task was difficult because the Office onet. pl employ 650 workers on the surface of a 4500m2.To functionally arrange space for conversations, meetings and work to come up with ideas,thethought of guiding, which incorporates all the aspects of the spread over 7 floors of an officebuilding and, at the same time, brand was Identified with the portal.

The inspiration to become the elements that surround us and at the same time are characterized byindividual divisions of the company. Uniqueness, lightness and panache of the proposed Interioracquaintances creating our own through the use of mainly non-finishing materials, furniture (mainlyPolish designers) not forgetting to design space has always been a functional and ergonomic.

First floor is brendingfloor –colors black and yellow logo as a background color of white. Next,refer to the color of these elements. And so we have fire, water, Earth and air. On each of thesefloors materials and their structure and colors are appropriate for the element .

On most floor ceilings are open, they are painted with the installations on the color of theelement. This solution has helped get more space and loft offices in nature where the majority ofworking young people

On the ground floor of an office building is a common space for employees Playroom.
A place where you can enjoy a cup of good coffee, sit down in a comfortable chair, put on thepicture taken in and even after swing on a swing, play Pinball machines but also in this field, youcan arrange a meeting.

Because if the worker to perform his duties, must sit behind the desk? Whether ridding it of anydistractions or conditions to rest will make will work more efficiently? I don't think so.Experience shows something quite the reverse. The longer you don't come away from the computer(often just there is too much where . . .), the harder it is for us to focus on. The longer weraise with the chairs, the more we are tired. The more similar to each other days spent at work,the less enthusiasm and desire to perform their duties.

An unusual interior has loosen rigid thought patterns and inspire employees to think of the custom.

NOMINEE FOR BEST DESIGN AWARD OF OFFIC SPACE
最佳办公空间设计奖提名奖

胡之乐（中国杭州）

获奖项目/Winning Project
浙江蓝城建筑设计有限公司办公室
Bluetown Architecture Office

设计说明/ Design Illustration

在只有5m多高的既有建筑空间内，排除设备高度后，如何构建一个建筑设计事务所，是本案的难度所在。在设计过程中和建筑师一起充分分析了空间之间的关系，结合人使用的方式，合理地规划高度，用联廊和玻璃盒子打造出一个简洁而空间丰富的场所，用体块关系在室内打造一个建筑是此案设计最初的构想。

How to create an architectural design studio within the five-meter high original building and take into account of the equipment height is the key difficulty of this case. In the process of design, we negotiated thoroughly with architects about the relation among spaces, people's habit of use, reasonable planning of space height, by integrating overhead corridor with glass box, we generated a simple and spatial rich place. The utilization of block relationship was the original concept of conducting the construction within this building.

185

IDEA-TOPS
艾特奖

获奖者/ The Winners
黄永才
（中国广州）

获奖项目/Winning Project
SONG'S 俱乐部/ SONG'S CLUB

190
IDEA-TOPS
艾特奖

获奖项目/Winning Project
SONG'S 俱乐部
SONG'S CLUB

设计说明/ Design Illustration

位于广州市CBD珠江新城兴盛路的SONG'S俱乐部是基于中国山水空间情趣以及时间与空间的叙事的想象所设计的一个集娱乐休闲与餐饮的功能场所。

SONG'S俱乐部由GRG所塑造的形体所形成的不同的围合并且互相渗透的空间是"宋"吧最显著的特点。GRG的工艺技术使得围合的墙体如山体般流动、起伏。我们从中国传统的山水画中寻找古人对空间、艺术、诗意的向往，用现代的语言创造出形体不一的起伏的"山墙"，围合出不一样的界面模糊暧昧的空间，垂直线条相交的装置墙体在灯光下透着光影，如自然的树影婆娑，粗糙的材料与光滑的金属贯穿不同元素，使原始与未来的因素在一个空间发生碰撞。

SONG'S俱乐部用了先进的GRG技术来建造空间里最有张力的"山墙"使得这些流畅起伏的像国画里的山体一样的造型能在SONG'S俱乐部这个空间里组合起来，形成富有艺术感的空间。在这些GRG的流质形体上贴上金属的质感，使得其更加有现代的质感与场所的属性。空间里除GRG外不乏一些手工工艺制作的艺术元素，垂直线条相交的装置与"song" logo的金属板，都是一些手工制作的艺术装置。现代与原始的因素在碰撞着。

亚洲有着自己独特的文脉，这些基因不会被忽略或者遗忘的。SONG'S俱乐部的设计呈现也基于对传统与现代的共性的思考的结果。东方对空间的理解，对诗意、自然、意境、对话的空间表达是有自己独特的领悟的。如园林的一步一景，开合的设置，漏透的呈现。我们都在尝试在现代的设计中寻找它的共性与延展。

SONG'S俱乐部在位于广州CBD这样的大都市的酒吧街中以自己独特的姿态获得非凡的回响，在众多酒吧中其差异性获得了广泛的关注，对细节及品质感的关注与执着使得SONG'S俱乐部的品牌效应正在起着作用！创新、品质已经和SONG'S俱乐部融合一起。这使其在激烈的竞争中赢得了不少的市场份额。

主要材料：GRG、铜片、电镀不锈钢、大理石、耐候钢、定做地毯、实木地板。

Settled in Xingsheng Road, Zhujiang New Town, a CBD in Guangzhou City, Song's Club is designed based on the charming chic of Chinese landscape painting and an imaginary narration between time and space, what endow the club the functions of entertainment and catering.

Entities shaped by GRG divide the interior layout of Song's Club into separate and interpenetrating spaces, which have become a beacon of Song's Club. The technology of GRG reconstructs the enclosing walls waving through the club like meandering hills. Inspired by traditional Chinese landscape painting what ancients opt for space, art and poetic sentiment, we adopt modern languages in creating uniquely contoured "gable walls" so as to bring forth an atmosphere of lingering in an indistinctively enclosed space, where the walls are decorated with vertically configured lines and with mottled lights shed shadow from above; rough materials staggering with smooth metals feature the club a place furnished with conflicting primitive and future elements.

The "gable walls" established by GRG technology wave through like meandering hills in Chinese painting, combine with the interior layout of Song's Club, bring out a space of aesthetic art. The liquid forms of GRG wrapped in metal cladding enhance the sensation of modern decoration and the natural property of the club. Besides GRG's products, there are some artificial handicrafts, assets with vertically configured lines and metal plate cast with the logo "song", all these assets are handmade. Song's club is a place where primitive conflict with future.

Asia has its very special cultures, some genes are never meant to be ignored or forget. The design of Song's Club is what brought out based on our perception of the commonness between traditional culture and modern culture. The orient has a unique way of understanding space and presenting poetic sentiment, nature, prospect and spatial display of dialogue. The scenery in garden, the open settings as well as the hollow-out and perspective displays, all these elements are embedded in modern designs to exhibit their commonness and extension.

Set in CBD in a metropolitan like Guangzhou, Song's Club earns its reputation as a spectacular symbol on the bar street, and has received great attention for its obvious differentiation, among which the emphasis and persistence on details and quality play a vital role in keeping the brand effects working! The fusion of innovation and quality has been part of Song's Club and helps it expand market share in the fierce competition.

KeyMaterials: GRG, copper, electroplating stainless steel, marble, weathering steel, customized carpet, solid wood floor

获奖评语

设计师营造了一个具有丰富想象力的、梦幻般的体验空间。

The designer has created a dream like experiencing space with rich imaginations.

IDEA-TOPS
艾特奖

NOMINEE FOR BEST DESIGN AWARD OF ENTERTAINMENT SPACE
最佳娱乐空间设计奖提名奖

李健彬&陈国斌
（中国佛山）

获奖项目/Winning Project
佛山市东方广场新地KTV
Newland KTV

设计说明/Design Illustration

本案以经济、快捷、有效为原则，将当地一个接近10年未改动过的老牌KTV，用最直观的"视觉标识"对其进行升级改造，重新注入活力，打破消费者对其形象的固有化认识。

并将这些"视觉标识"以"安装"的手段解决了业主对项目的分区域且不停业改造，不对整体重新布局，每个改造区域工期短等要求。

This project is supposed to be economic, quick and effective in principle. It will transform and upgrade a local veteran KTV unaltered for 10 years with the most intuitive "visual identification", revitalize it, and break the consumers' understanding of inherent image. z
And the "visual identification" will solve the problems that the Entertainment not closed during the transformation, no overall layout renewal and short-term transformation areas by means of "installation".

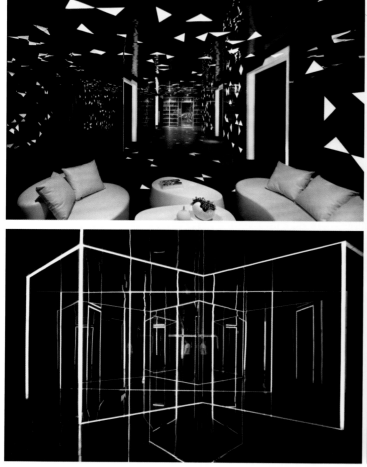

NOMINEE FOR BEST DESIGN AWARD OF ENTERTAINMENT SPACE
最佳娱乐空间设计奖提名奖

裘小松（中国成都）

获奖项目/Winning Project

兰桂坊fervor酒吧
Fervor Bar

设计说明/ Design Illustration

项目地址位于成都水津街兰桂坊8号，周边酒吧及餐饮云集。周边的酒吧多为慢摇或演艺吧，装修风格偏富丽豪华，大量运用水晶、石材、不锈钢，如何让本项目脱颖而出，吸引众多玩家及泡吧族追捧，设计只有另辟蹊径。

经过与甲方多次沟通，最终选择以太空为主题，打造科幻风格。酒吧入口设计成飞船太空舱大门的造型，配合两侧大面LED画面，从门外就开始给人即将进入科幻大片的感觉。室内利用层高优势，由中心散座区向两侧逐级抬升，各层平台均为流线型层层叠加，如同云层般的感觉，并利用较高的层次将库房、操作间及卫生间均设置到卡座区的下方，最大限度的提高了空间利用率，同时使整个室内空间没有视觉死角。酒吧的墙面都使用白色GRG造型，各种灯光投射在流线型的白色墙面及栏杆、网架上，幻变出各种色彩，如同打开了潘多拉的魔盒。整个大堂顶棚由整块的LED屏构成，并一直延伸到吧台背后的墙面，各种主题的素材不停播放，时刻给人以太空穿梭的感觉。作为动静分区，酒吧内还设计了两个如同太空舱般的包间，可容纳20~30人，专为派对或主题活动准备。
整个酒吧空间光影流动，如梦如幻。

Located at No.8 LanguifangShuijing Street in Chengdu City, the project is surrounded by numerous bars and restaurants. The bars nearby are mainly slow wave clubs or performing bars, decorated with luxurious styles, using a large number of crystal, stone and stainless steel. How to grant the project a provocative superiority and become a boom among trendy players and club fans? Only a surprising design can be counted on.

After negotiating with Party A for several times, the interior design of Fervor Bar finally chose the theme of Space, to endow with a science fiction style. The architecture design applies a spacecraft capsule at the entrance and large LED displays on both sides, makes you feel like stepping in a fictional movie scene. Taking advantage of storey height, the stage uplifts from the center floor area to both sides, terraces pile up like streamlined clouds, layer upon layer; considering the height of layers, we allocate warehouse, operating room and restroom underneath the deck area so as to maximally improve space utilization and avoid blind corners inside the bar. The walls are decorated with white GRG entities, a collection of colorful lights project on the streamlined white walls, handrails and wire frames, make this bar a Pandora's Box. The hall ceiling is composed of a whole LED screen, extending to the back wall of the bar and playing different themes of videos to bring forth an atmosphere of space shuttle. Two capsule compartments, which function as living and activity areas, can accommodate 20-30 people to conduct parties or theme activities.

The whole bar projects colorful lights like a magical fairyland.

IDEA-TOPS
艾特奖

NOMINEE FOR BEST DESIGN AWARD OF ENTERTAINMENT SPACE

最佳娱乐空间设计奖提名奖

Miguel de la Torre Arquitectos
（墨西哥）

获奖项目/Winning Project

光之魅
Árbol CDMX

设计说明/ Design Illustration

在这个当地居民和参观者欢呼的纪念性空间里，一个可观的礼物、一缕阳光和良好的祝愿就是墨西哥城给予其公民欢度2014年圣诞节的礼物。该城市希望，不仅采用这种华丽的姿态来奖励，而且还带着大家在一个适合共存的多彩和点亮的空间里一起欢乐。

该"Árbol CDMX"的装配不同于任何以往建成的用以庆祝冬天假期的建筑。需要注意的是，它是由一群墨西哥创意者在墨西哥进行思索、开发和制作出来的一个理念。这一建筑体不同于常规的圣诞树；它是一个立方体建筑，象征着具有巨型窗口以及"季节快乐"的标志性人物形状的礼盒。

该创意团队是由建筑师Miguel de la Torre和设计者Jorge Cejudo组建而成的，这种盛大的树不是用于沉思，而是邀请参观者通过光线、音频和视频节目进入并经历一个独特的体验。在白天，参观者享受了一种圣诞节的夜晚、节日，而且在其室内充满着光线。当太阳下山后，这种巨大的礼盒是被来自四方的互动光线覆盖着，因而增强了其可见度，同时模拟一个灯塔，邀请所有的历史中心步行者都加入到这个空间里。在这一天结束的时候，在音乐和灯光的环绕下，向参观者演示体验一种象征性的以及非常有吸引力的人工降雪。

冬季不仅仅有圣诞节；这是一个喜庆的时刻，此时人们关闭循环、下定决心，并且充满爱和希望。在这样一个充满礼物、祝福、家庭、团聚、光线和色彩的时刻，有这样一个理念，基于给居民光线并希望他们用自己的光线将其充满。这个项目总结了光线显示的途径，这是快乐的，因为这消除了黑暗，同时给予了我们一个积极的参考。正是这个光线，我们大家作为公民都需给予和分享。该树是那个空间的一个寓言，我们所有墨西哥人（当地居民和参观者）充满着自己的光线。

A grand gift, a tree of light and good wishes is the present that Mexico City gave for Christmas of 2014 for its citizens in this monumental space where locals and visitors greet. The city wanted, with this magnificent gesture not only to reward, but to bring together everybody with joy in a colorful and lighted space ideal for coexistence.

The Árbol CDMX is an installation very different from anything that has been done to celebrate the winter holidays. It is important to note that it is a concept that was thought, developed and produced in Mexico by a group of Mexican creatives. This piece is very different from the regular Christmas trees; it is a cube that symbolizes the gift box with enormous windows with the shape of this iconic figure of the "season to be jolly."

For the creative team, formed by architect Miguel de la Torre and designer Jorge Cejudo, this grand tree is not for contemplation, it invites the visitors to enter and live a unique experience through light, audio and video show. During the day the visitors enjoyed a Christmas night, festive and full of light in its interior. When the sun went down this huge gift box is covered with interactive light in the four sides, increasing its visibility and emulating a light house that invites all the historic center walkers to join together in this space. At the end of the day show the visitors experience a symbolic and very attractive artificial snowfall encompassed by music and light.

The winter season is not just Christmas; it is a festive time when people close cycles, make resolutions and fill with love and hope. A time for gifts, wishes, family, union, light and colors with a concept based in giving light to the inhabitants and hoping they will fill it with their own light. This project summarized the light that shows the way, which is joy, which removes the darkness and gives us a positive reference. It is this light that we all as citizens have to give and share. The tree is an allegory of that space we all Mexicans —locals and visitors— fill with our own light.

NOMINEE FOR BEST DESIGN AWARD OF ENTERTAINMENT SPACE
最佳娱乐空间设计奖提名奖

蒋国兴（中国深圳）

获奖项目/Winning Project

深海壹号餐厅酒吧
Ocean Restaurant Bar

设计说明/ Design Illustration

本案位于新疆乌鲁木齐南湖东路酒吧园区，是近年新兴的，聚集娱乐餐饮为一体的商业圈。深海壹号主营海鲜火锅，是一家以海洋为主题的高端餐厅。

通过与业主沟通，设计师对海洋概念的了解更深，形成一个用概念手法诠释海洋文化的主题餐厅的设计。

不同的餐饮文化主题给人不同的感受，以海洋为主题的餐厅蕴含着一种蓝色的氛围。蓝色调体现着丰富的情感内涵，蓝色让人联想到广阔、深远。波浪滔滔的大海忧郁的情感总是与蓝色连在一起，使人感到优雅、宁静。金色是餐厅的第二色调，主要分布在餐厅顶部。顶部保留原结构造型，增加线条装饰，配以暖色灯光造成产生光明、华丽、辉煌的视觉效果，金色宛如轻柔的沙滩，沙滩与海相辅相成，海浪静悄悄的通过，又悄悄的退去，聆听着海风，感受沙滩的柔软、温暖、散发太阳的气息，享受着美好时光。

大厅入口，映入眼帘的是一艘做旧大型古帆船，经过帆船你会观赏到船头礁石上美人鱼的动人姿态，布局上大厅动线流畅，没有阻挡，而地面则采用各色条型大理石铺设，增加大厅的层次感，犹如海浪般让人觉得自己不是在走路，而是驾着一艘舰艇穿行在辽阔无际的海洋，自由而又惬意。帆船的上方，悬挂成群的银鳞小鱼，如有生命的鱼儿嬉戏追逐，成群结队，让人更加亲近自然、热爱生活。海浪、沙滩、鱼群、古船、鸣笛、深海生物造型的纸灯还有老船长，这些都让你不禁勾起童年的回忆。大厅右侧是服务台区，服务台前部由白炽灯泡装饰，配上蓝色灯光，流光溢彩，有种梦幻质感。左侧设置了一个等候区，圆形的卡座沙发融入环境，休闲舒适，在餐前来客可以在这里小憩一会儿，可以与友人聊聊天，也可以欣赏餐厅主人为你安排的音乐节目。再往两侧去是餐厅的散座区，散座以鱼鳞状隔断分隔，蓝色的麻布沙发，充满"海洋"的气息。而开放散包用灯泡装饰隔墙，宛如深海的泡沫向上漂浮，配上蓝色灯光营造的氛围，浪漫迷人。蓝白的椅子增加此空间的活跃感，再加上精心挑选的餐具，这些共同营造出高尚雅致的用餐环境。包厢部分集中在餐厅二楼，大包厢的顶部延续了大厅的元素，保留原结构以金箔贴顶加以石膏线条，蓝色的壁纸墙面，突显海洋气息，并将活泼好动的海洋生物饰品融入环境当中，在安静的空间内增添了趣味性。包厢并不是全封闭性，在包厢靠近大厅的侧面墙，设计师运用鱼鳞隔断预留位置分隔，在来客用餐的同时也可以欣赏大厅的音乐节目，丰富用餐乐趣，把周围的环境应用与调动起来，充满人文气息的知性风格，融入了人们对生活的品味和期许，优雅舒适。

现代人在品味美味佳肴的时候，开始关注用餐环境的文化氛围和个性化，而餐厅的设计正是满足顾客这部分需求。在这个像大海一样不可测度的魅力餐厅，宁静、神秘、新奇，在身临其中的时候，通过视听与联想，希望能让人进入期望的主题情境。

Located at the bar park of Nanhu East Road, Urumqi City, Xinagjiang, Shenhai Yihao is a high-end restaurant specialized in seafood hotpot. Upon in-depth communications between the owner and the designer, a better understanding of seafood is achieved to interpret the restaurant themed with seafood culture.

Different culinary culture themes bring in different feelings for people. The restaurant themed with seafood contains a blue tone which embodies rich emotional connotation and reminds people of the broad, far-reaching and blue sky. Besides, the gloomy emotion of the choppy sea is always associated with blue, which makes people feel elegance and tranquility. Gold is the second color tone which is mainly embodied on the top of the restaurant. The top maintains the original structure modeling and adds line decoration, coupled with warm lights to have the bright, gorgeous and glorious visual effect. Gold is like a gentle and soft beach, while the beach is supplementary to the sea. When sea waves come and go quietly, we can feel sea breezes and enjoy the softness of the beach and the warmth of the sun to have a good time.

At the entrance of the hall, you will see a distressed large ancient ship. By passing by the ship, you will see the charming posture of the mermaid on a rock on the ship's bow. In layout, the hall features dynamic lines without hindrance. The floor is paved with strip-type marbles of various colors, which makes the floor more layered, as if people are driving a vessel on the vast ocean freely and cozily. A group of small silver-scale fish is hung on the top of the ship, as if living fish is chasing and romping in groups, which makes people close the nature and keen on life. The sea wave, beach, fish school, ancient ship, whistle, deep-sea organism paper lantern and old captain make you memorize your childhood. On the right of the hall is the service counter area. The front of the service counter is decorated with incandescent bulbs accompanied by blue lights, so gleaming that a dreamlike texture is fostered. On the left, a waiting area is set with round booth sofas which fit the environment for the rest of guests. Further to the two sides are the randomly placed seat area and open randomly placed seat boxes. The randomly placed seats are separated by scale-type partitions, equipped with linen sofas and full of the breath of "ocean". The open randomly placed seat boxes adopt bulk for decoration of partitions, just like foams floating upwards on the deep sea, so romantic and charming with the amenities of blue lights. Blue and white

chairs enhance the vividness of the space. In addition, the tableware is meticulously selected. All these develop a noble and elegant dining environment. Boxes are set on the second floor. The top of large boxes also use the elements of the hall, while the original structure of gold foil and gypsum line is retained. The blue wallpaper highlights the breath of ocean and integrates lively marine organism accessories into the environment, which adds interests for the tranquil space. The boxes are not totally closed. On the side wall next to the hall of the boxes, the designer uses scale-type partitions for place reservation, so that guests can enjoy the music program of the hall at dining. Such a elegant and comfortable surrounding environment is full of humanism that integrates the taste and expectation for life of people.

When enjoying delicious foods, modern people tend to focus on the cultural atmosphere and individuation of the dining environment. The design of the restaurant is just to meet such demands of guests. In this charming restaurant like ocean beyond measurement, you will feel its tranquility, mystery and novelty. Furthermore, you can enter the theme context that you like via seeing, hearing and association.

212
Best Design Award
Of Dining Space
最佳餐饮空间设计奖

艾特奖
最佳餐饮空间设计奖
BEST DESIGN AWARD
OF DINING SPACE

IDEA TOPS
INTERNATIONAL SPACE DESIGN AWARD

获奖者/ The Winners
徐梁
（中国杭州）

获奖项目/Winning Project
海盗鲜生/ Pirate Seafood

214

IDEA-TOPS
艾特奖

获奖项目/Winning Project
海盗鲜生
Pirate Seafood

设计说明/ Design Illustration

此案空间是餐厅、是酒吧,是年轻人互动社交的平台,也是个性、叛逆、不安现状、颠覆传统的生活方式。其间格局、装饰粗旷赤条但又不乏铺陈有序,与餐厅主创的海鲜裸烹却又不失美味遥相呼应。

空间最初给人一种"牢笼"的意象,如同束缚海盗自由之身的牢狱,隐喻传统思想与平淡点线生活对现代人追求新鲜、拒绝平凡的强烈束缚与冲突,"牢笼"间的围合,游走于钢网建筑与透视之间,鉴于空间内象征与意象的性质,该意象作为主要装饰元素被运用于整个空间,同时也从视觉上改变着空间的规模,任何人置身于此空间,便可切身感受到自己既是这个时代传统势力、思想的被束缚者,也是敢于冒险勇于尝鲜追求自由快乐的发起人与倡导者,更是偷得浮生,乐享鲜活的传染者。

层层黑色钢网矗立在整个空间,粗糙而原生态的水泥墙面与随处可见的骷髅,无一不在提醒着过往的来客:人生短暂,不论生活带给了你多少黑暗与碰壁,也不要辜负青春与光阴,尽情地敞开心扉去表达、去交流、去释放,每一天都要过得鲜活而精彩!

As a restaurant & a bar, Pirate Seafood aims to be a social platform for youngsters, expresses its attitudes toward personality, recklessness and spirit of subverting tradition. The interior layout and decoration endowed with a rough style in an ordered way to echo the spectacular healthy seafood cuisine.

The bar may first leave you an impression of lingering in a prison, where restricts the spirit of freedom in a pirate, reflecting strong bound and conflicts existing between tradition and nowadays plain routine life, and implying the hidden desire of making a break through. Enclosures, whether wire configured or perspective, considering the properties of symbolization and image, shall be applied as decorating elements to the entire space, so as to transform the visual space size, anyone who wanders inside can clearly grab the feeling of being restricted whilst withdrawing into a reckless adventuresome situation, rendering in a radiant disseminator of passion and enthusiasm.

Black layers of steel nets erect towering in the bar, coarse original cement walls and skeletons are visible everywhere, reminding the visitors that life is short, no matter how much darkness and refusal you have been undergone, do not waste youth and time, keep an open mind to express, communicate and release, embrace every fresh and wonderful moment!

IDEA-TOPS
艾特奖

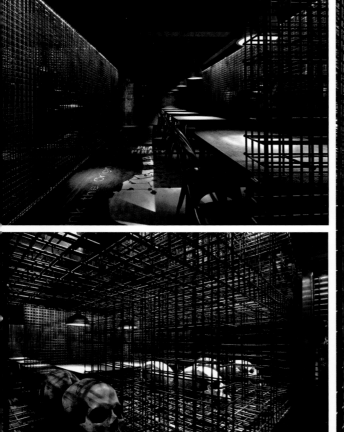

获奖评语

戏剧化而偶像化、碎片化而统一化，该餐馆让客户惊诧又大饱眼福。

Dramatic but iconic, fragmental but unified. This restaurant in a surprise for the customer and a joy for the eyes.

NOMINEE FOR BEST DESIGN AWARD OF DINING SPACE
最佳餐饮空间设计奖提名奖

Ion Popusoi（罗马尼亚）

获奖项目/Winning Project

雪山餐厅
Restaurant Gradinita

设计说明/ Design Illustration

餐厅的构思是森林中的一间小房子，旨在将餐厅本身融合到景观当中。为了维持原貌，设计时将现有树木融入了建筑结构之中。
该建筑包括两层：下层为透明；上层为不透明，具有装饰性。
下层全部安装玻璃，带来极大的透明度。外部空间可作为内部空间的延伸，同时与外部景观融为一体。建筑物立面由滑动表面构成，几乎可全面释放内部空间。
室内装修并没有受建筑物的外部限制，而是通过玻璃幕墙向前迈进，进一步消除了室内外的界限。
室外平台所种的植物是下一道视觉界限。
上层不透明的阁楼覆盖着大型的空间结构，呈独立的一层。是一个介于人造与自然之间的空间，也是室内与室外风景的过渡。
该结构以程式化的方式，通过一个花园与建筑物连接，灵感来自于自然的多样性。阁楼优化了现场地形，采用强烈的雕塑特征将建筑物转变为抽象体。创造出一种人造景观悬于绿色空间上方的错觉。凭借多层次重叠规划，建筑物立面获得了额外的尺寸和深度。垂直金属元件的基座种植了常春藤，覆盖了部分空间结构，使人造地形与自然地形之间相得益彰。
这栋建筑物会随着气侯的变化而变化。

The restaurant, conceived as a house in the woods, seeks to blend itself in the landscape. In order to leave the natural evironment unchanged the existing trees were included into the façade structure of the building.

The architecture consists of two layers, the lower – transparent and the upper – opaque and ornamental.

The lower register, fully glazed is dematerialized through transparency. The outer space can be captured as an extension of the interior space and the architecture in time will begin to blend with outer space in a landscape continuum. The facade consists of sliding surfaces that open interior space almost completely full .

Interior finish does not stop at the outer limit of the building, advancing through the glass façade , further reducing the distinct boundary between inside and outside.

Next visual boundary is the vegetation planted on the terrace outside.

In the upper register, opaque attic is covered by a large spatial structure , an autonomous layer, a space between artificial and natural, also a transition between interior and exterior scenery.
This structure connects the building with the garden in a stylized way, inspired by the natural diversity. The attic completes the site topography, transforming the building into an abstract body with strong sculptural character. It creates the illusion of an artificial landscape suspended, levitating above the green space. Building facade acquires an extra dimension and depth due to multiple overlapping plans. At the base of the vertical metal elements was planted ivy, covering partially the spatial structure , to complete symbiosis between the man made topography and the natural topography.

Following the climate this layered building will be in constant mutation.

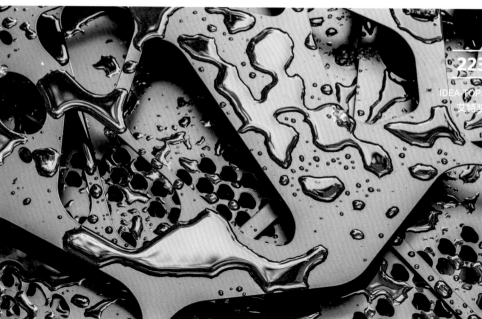

NOMINEE FOR BEST DESIGN AWARD OF DINING SPACE
最佳餐饮空间设计奖提名奖

黄鑫（中国南昌）

获奖项目/Winning Project

南昌佐祐餐厅凯德店
City Forest

设计说明/ Design Illustration

本案设计师在设计佐祐餐厅之前，思考着一个问题：生活在城市的人们，应该过一种什么样的生活？在这个空间里，从餐饮角度思考人们需要一个怎样的城市；生活在车水马龙间的人们，如何享受生活。

如果我们生活在这样一个靠海不远的城市，四周不是花田就是水稻田，住的是老房子，房子就在田中间，没有围墙，只有一丛丛竹子将人们隐在里面。佐祐还原的便是这一生活场景，迎宾接待区，为茅草屋，身后是一片绿色，簇拥的竹子之间，设计师巧妙地将竹子的镂空用来培植其他植物，既节约用地，也美化空间。

来到这座城市，在柏油路上行走，我们随眼可见遮天蔽日的树木，参差不齐，遮挡了夏天的烈日，挡住了冬日的寒风，树木在粗糙的城墙倔强生长。故用水浇泥、鹅卵石、岩石构造墙面，回廊之间用水泥铺就，黄色的斑马线，映射城市繁华。

知了、昆虫、鸟在身边叽喳不停，这个时候你会取下耳机，听它们在为这个城市演唱。设计师在每个餐饮区和树木上都进行细致处理，契合城市森林生态链。走累了，就在书吧歇脚，空间搭造一座方形书屋，用白色书围护空间隐蔽性。

每当夜幕降临，穿过萤火虫之森，投影技术折射的光点，强化萤火虫发光效果。手绘的两旁青树，取自动漫作品《萤火虫之森》灵感，讲述一人一妖的爱情故事，因为偶然机会相遇，偶然的机会相拥、相别。他们没有轰轰烈烈的相遇，却有促不及防的分别。犹如设计师将此设为隧道状，并取名为"爱情隧道"。

通过萤火虫之森，城市和乡村的距离对望，能够望见乡村院落里，枝桠爬上窗台，麻雀每天早晨在阳台将晚起的人叫醒，墙根下，松鼠打转巴巴看着阳台上快要垂下的果子。这样的情景，设计师表达在他的画中，一颗大树从座位中伸展开，随着袅袅炊烟，往上长。

我们居住的城市是这样一个大森林，清晨小孩子们跑去竹林挖竹笋，午间我们端着凳子，摆上桌子，在树下吃饭、乘凉；下午，跑去书店，翻看闲书，或呆在家里听音乐，晚间捉萤火虫。如果到了夏天，我们搬出凉席，和邻居的小孩一起数星星，一边数，一边睡着。爷爷奶奶拿个蒲扇慢慢摇，让我们安睡。

这就是城市生活，让车水马龙给生态腾出空间，有蓝天、青砖、灰瓦，不紧不慢穿行在老街坊长长的光影间，路人礼貌客气、干净体面，过着简单而讲究的日子。

Before the design of Zuoyou Restaurant got started, the designer in charge of the case couldn't help thinking about a question, to people living in the city, what kind of life do they deserve? In this building, we can notice that, from the angle of catering, the thinking about what kind of citywe need, living among the hasty crowd, how to enjoy our lives?

Suppose we live in a city not far away from sea, surrounded by flowers and paddy fields, in an old house erects right in the middle of the fields, fenceless, only bamboo forests conceal us inside.What Zuoyou restores is this kind of living scenario, it decorates reception as banda and uses bright green as its background, among the lush bamboo forests, the designer subtly inputs some hollowed-out bamboo plants to save space and beautify environment.

When you come to a city and wander along asphalt road,what you want to see islush trees bordered on both sides of street,towering over head, blocking the scorching sun in summer and piercing wind in winter, and taking root beside rough city walls. Therefore, I poured water to cement, used pebbles and rocks to decorate the walls, paved the cement and yellow zebra between corridors to reflect the bustling city.

Cicada, insects and birds twittering here and there, take down your earphone, listen to them singing. The designer carried on subtle treatment to each dining area and tree to better echo the ecological chain in Urban Forest. After a tiring day, walk in and relax yourself in a square bookstore where privacy is protected by white enclosure.

In the evening, when night falls, wander through the forest of fireflies to experience the glowing effect of fireflies and refraction of light created by projection technology. Inspired by the cartoon the Forest of Fireflies, painted trees on both sides to tell a fairy tale about a man and a goblin who met each other by chance, embraced accidently and separated. They had no dramatic encountering but a hasty farewell. Thus the designer furnished here as a tunnel and named it "Love Tunnel".

The city and country locate attwo sides of forest of fireflies, from here you can see clearly the courtyard in countryside, branches stretching to the window, sparrows singing and waking people up in the morning, squirrel eagerly staring at the ripening fruits. The designer expressed this scene in his painting, a big tree spreading its branches among the seats and curling up among rising smoke.

The city we live in is a grand forest where children can dig bamboo shoots in the morning, grown-ups lay the table and have meals undershadow of tree, spend the afternoon at bookstore or stay at home listening to music, then catch fireflies in the evening. In summer, we can pull the mat outside and count the stars with kids while grandparents sitting around and enjoying this sweet moment.

This is urban life, clear a room from the hustle and bustle, enjoy the blue sky, white wall and grey brick, spend your time with old neighbors, everyone treats people politely andlives a simple and elegant life.

NOMINEE FOR BEST DESIGN AWARD OF DINING SPACE
最佳餐饮空间设计奖提名奖

Miguel de la Torre Arquitectos
(墨西哥)

获奖项目/Winning Project

传奇
La Legendaria

设计说明/ Design Illustration

墨西哥城的市中心区提供了各种各样的选择，因此，对每个人来说都有一些机遇。近些年来，该城市的这种传统区域的印象已经发生了改变，因其优先考虑到行人并且建立起了一种新的氛围。商业活动已经由于这种转变而获益，与之相应的是，出现了各种新的美食选择，并在食物、设计和服务之间趋于完美的平衡。

本案位于乌拉圭大街，其完美地体现了这一理念，敞开着大门，给人这样一种印象：向你发出邀请，让你走进里面，会有更多发现。该项工程完工于一座宏伟建筑的第一层楼，而该建筑曾经是该城市的金融区，展现了一种具有吸引力的现代氛围，以满足当地居民和参观者的口味。

该项工程是由Miguel de la Torre Arquitectos团队予以完成的，该团队以设计理念的个性为主轴，并将传统墨西哥菜肴中的木制勺作为设计的主要元素。重复性的元素是建筑师Miguel de la Torre所有工程项目的一个特征，而在这一特定项目中，工程进行得很好，因为任何食谱（特别是该餐馆的特色菜mole negro）做到尽善尽美的唯一方式就是一遍又一遍的重复，直至熟练掌握，这一细节到处都可以欣赏到。

该房屋面积350m²，其中200m²用于客户服务，而其余的面积用于厨房及相应的服务设施。该餐馆完全面向大街，充分利用了繁忙的客流特别是白天人流量更大这一特点。其主色调是黑色的，而在其右侧墙壁上，从地板到天花板，装配了一万个木制勺，从而创立了这样一个特色，不仅使其具有动感，而且给墨西哥菜肴增添了这种实质性天然材料的温馨感。

该地区分为三个大的区域。休息室位于前台，在一个玻璃箱内设有吸烟区，主客厅在后面，最后是一个连接两个空间的30m长的酒吧。恰当的利用该房屋的尺寸面积，让客人感觉到一个很宽阔、大气的空间，之所以取得如此成效，是由于主要通道和走廊的位置是从入口直到末端。厨房设在后台，而其部分面向客厅开放，以便让客人感受其操作。

Mexico City's downtown offers a great variety of choices, there is something for everyone. In recent years the image of this traditional area of the city has changed giving priority to the pedestrians and creating a new atmosphere. The commercial activity has been beneficiated from this transformation and to be consistent new gastronomic choices have emerged with perfect balance between food, design and service.

La Legendaria - Gastrocantina- located in Uruguay street, understood perfectly this concept opening its doors with an image that sends an invitation to go inside an discover it. The project was done in the ground floor of one of the magnificent building, which once was the financial district of the city, to present an attractive contemporary atmosphere to satisfy the taste of locals and visitors.
La Legendaria - Gastrocantina–The project was done by Miguel de la Torre Arquitectos team who choose as the main axis and personality of the design concept an element of the traditional Mexican cuisine: the wooden scoop. Repetitive elements are a characteristic of the projects of architect Miguel de la Torre and in this particular project it worked very well because the only way that any recipe - especially mole negro, the house specialty - is done to perfection is to repeat it over and over until mastering it, a detail that can be appreciated all over the place.

The property has 350 sq m, 200 were used for customer service and the rest for the kitchen and services. The restaurant is completely open to the street to take advantage of the heavy traffic especially during the day. The main color is black and on the right side wall 10 thousand wooden scoops were installed from floor to ceiling creating a texture that besides being dynamic adds the warmth of this essential natural material in the Mexican cuisine.

The area is divided into three big zones. A lounge in the front inside a glass box that generates the smoking area, the main salon in the back and finally a 30 m long bar connecting both spaces. The correct use of the dimensions of the property allows the customer to feel in a very wide and large space achieved by the location of the main access and the corridor that goes from the entrance to the end. The kitchen is in the background and partially open to the salon to perceive the operation.

NOMINEE FOR BEST DESIGN AWARD OF DINING SPACE
最佳餐饮空间设计奖提名奖

徐代恒（中国南宁）

获奖项目/Winning Project

优鲜馆(万象城店)
FreshT

设计说明/ Design Illustration

空中洋溢着新鲜果香，四处弥漫着原始木色，既温柔又精致的优鲜馆果汁店无时无刻不展现出其特有的态度与腔调：以轻奢品迎新贵客。

低调的砖墙作为墙裙与"青石板路"呼应延伸，巧妙地将中古世纪的欧洲集市带到了人们眼前，供匆忙的现代人往返流连，细细回味。

拓展视线的镜面削弱了空间的束缚感，柜台之上升起的钢化玻璃让气氛变得轻盈，不加修饰的方柱又让一切回归平衡、舒服、明朗。

冷酷的黑铁层架如同黑色琴键一样穿梭在令人感觉温暖的水曲柳实木板中，一静一动，清新而不浮夸，热闹却不喧哗，还能将鲜艳的水果衬托得更秀色可餐、香郁诱人。

而将一切元素轻轻融合的是十来盏柔光夺目的玻璃吊灯，过渡了冷的铁，点缀了暖的木，让原本并不宽大的空间显得宽敞明亮，让整个设计变得浑然天成。

这间高调与内敛共进的优鲜馆，这间新鲜与沉淀共存的小集市，和其中满载着诚意与创意的美味果汁，定能让你暂时停下匆促的脚步，获得一刻完美的享受。

现代人在品味美味佳肴的时候，开始关注用餐环境的文化氛围和个性化，而餐厅的设计正是满足顾客这部分需求。在这个像大海一样不可测度的魅力餐厅，宁静、神秘、新奇，在身临其中的时候，通过视听与联想，希望能让人进入期望的主题情境。

With the aroma of the freshest fruits in the air, FreshT always demonstrates its great glamour. A gentle, exquisite and light luxury atmosphere penetrates from the pure wooden color in every corner, and delivers FreshT's positive attitude and uniqueness to the customers.

The understated brick wall, along with the wainscot, poses a "bluestone road" around the room. Such design portrays a vivid European fair from the Medieval Era, and offers the busy modern people an opportunity to travel back in time, with a long and enchanting aftertaste.

The mirror in the room weakens the limit of the space and expands the sight. The tempered glass above the counter raises the light and relax atmosphere. The square cylinder without the decoration brings back the cozy and comfortable balance.

Cold and black steel shelves stroll through the solid ashtree wood, just like the black keys in the piano. When static and dynamic objects support each other, it modes fresh with no flippancy; it brings fun with no noisy. Mostly, it instigates the taste and the charm of the fruits.

Quietly and softly, a dozen of the glass lights mix the whole world together. The light burnishes the cold steel, and flares the warm wood. Such nature-made design conjures up a bigger, harmonious and glossy macrocosm.

With both extraordinary and introverted characteristics in this little fair, Fresh T assembles fashion and classic, and composes its irreplaceable fruit products with good faith and endless creativity. It will for sure to slow your steps, unload your burdens, and regain the pleasure.

When enjoying delicious foods, modern people tend to focus on the cultural atmosphere and individuation of the dining environment. The design of the restaurant is just to meet such demands of guests. In this charming restaurant like ocean beyond measurement, you will feel its tranquility, mystery and novelty. Furthermore, you can enter the theme context that you like via seeing, hearing and association.

IDEA-TOPS
艾特奖

235

IDEA-TOPS
艾特奖

236
Best Design Award
Of Exhibition Space
最佳展示空间设计奖

艾特奖

最佳展示空间设计奖
BEST DESIGN AWARD
OF Exhibition SPACE

IDEA TOPS

INTERNATIONAL SPACE DESIGN AWARD

237
IDEA-TOPS
艾特奖

获奖者/The Winners
广州市形而上装饰工程有限公司
（中国广州）

获奖项目/Winning Project
赛德斯邦/Cerlords

获奖项目/Winning Project
赛德斯邦
Cerlords

设计说明/ Design Illustration

主题: 术业有"砖"攻
一切从廊说起……

横向: 廊是开放开阔的空间,廊,移步异景,包括的是廊还有廊之外的风景;

纵向: 廊是曲径通幽,有着无限延伸的可能和神秘。

让惯性思维停止!
在此改变的不仅仅是平面、空间的格局和形态,而是打破一种惯性思维,把平常的事物进行了异化,由此产生距离感和陌生感,从而最终改变了人与物的关系:重新的认识和价值的重估。

Design Specification of Cerlods - Flagship Store at Headquarters
Theme: Specializing in Bricks
Everything starts from corridor:
Horizontal: as an open space, sceneries vary while wandering along the corridor, inside or outside;
Vertical: corridor is a meandering path leading to secluded scenes, brings forth unlimited possibilities and mystery
Stop inertial thinking!
We are capable of changing more than spatial layout, structure and shape
To break through inertial thinking, and pursue for heterization of the ordinary
Sense of distance and strangeness will change the relationship between human and substance: redefine perception and reevaluate value

获奖评语

成熟的材料运用，丰富的空间形体。
Mature application of materials, rich spherical bodies.

NOMINEE FOR BEST DESIGN AWARD OF EXHIBITION SPACE
最佳展示空间设计奖提名奖

翁瑞栋（中国杭州）

获奖项目/Winning Project

金隅学府售楼部
Jinyu Bibliotheca

设计说明/ Design Illustration

纵贯古今横扫中外
古人云：耳读书而聪，目读书而明，心读书而一，神读书而注，凝读书而遍，虑读书而莹，饥读书而饱，困读书而醒，愠读书而吉，愤读书而平，噫，"余白首未闻道兮，唯读书以毕此生。"学无止境，路漫漫其修远兮，吾将上下而求索。金隅学府座落钱江新城核心区，周边豪宅楼盘林立，在延续豪宅定位的同时，项目强调"当代屋檐下的明媚家风"的教育文化主题，力求打造一个全新的高品质楼盘。
本案是金隅学府楼盘配套售楼部。为契合学府的主题思想，并针对目标客户群所具有的开阔的国际视野及深厚的文化底蕴，通过设计师兼收并蓄地匠心打造，一个纵贯古今、横扫中外的超级图书馆式的学府售楼部横空出世。

售楼部前厅
借鉴全球知名的"大英图书馆"中穹顶风格，设计出浓郁风情的书架，风格奢华大气而又不失雅致。英国图书馆是一座创造性、资源性、高效率的图书馆，采用这样的设计不仅仅是为了体现外在的华美，更重要的是以此体现出学府所拥有的创造性、资源性的内涵。

复古私塾
通过设计师的精确考证，1:1打造出经典的复古私塾，宛如穿越时空，来到鲁迅先生笔下的"三味书屋"，让我们更真切地体验到传统文化的魅力。睿智如你，可以坐拥书城，跟紧时代的步伐，颠覆你的人生观；也可以与更杰出的人物相识，为明天谱写更辉煌的篇章。

上传统意义的室内风格，而是从容地揉合了不同的元素，整体呈现出自由混搭的折衷主义风格。目光所及之处，精彩灵动而来。漫步在高挑的欧式圆拱门型通道，如同进入知识与智慧的殿堂接受神圣之洗礼，充满了高贵与庄严感；通道地砖的图案与顶天立地的镂空金属玄关遥呼相应，在光与影的交织下融为一体，幻似一条通往智慧的天堂之路。大面积烟雨灰的背景墙渲染出中国写意书画之美，配合古风十足的茶具、书画和瓷器等鉴赏品，勾勒出一派典雅的中国民族风。
在空间上涵盖了较多的功能，除了传统售楼所需要的功能区域，如咖啡吧、鉴赏区、洽谈区等，还有史无前列的图书馆阅览室。
特别值得一提的是运用传统与现代相结合的手法打造的鉴赏区。现代风的沙发，可以安放疲惫的身躯，令人舒适放松；纯原木打造的。才艺展示区由中国书房之文房四宝坐镇，高雅人士雅集于此正是切磋技艺之绝佳之地；在同样由纯原木打造的古朴风的品茶区。
煮一壶秋水，品一杯春茶，叙一怀夏情，默默安享时光静好！一个优秀的鉴赏区让人品味的是一种生活和体验！
史无前列的图书馆阅览室仿造大英图书馆的风格，特别的高大上，把时光留在书籍和智慧中，是人生最明智的选择！

UnprecedentedEver or Never
Foreword
As the old saying goes, reading with your ear can sharpen your listening, with your eye can expand your perspective, with your heart can concentrate your mind, with your soulcan calm yourspirit, reading with attention so as to digest, with contemplation so as to reason, with hunger so as to stuff, with confusion so as to sober up, with anger so as to delight, with resent so as to pacify, alas,"White-headed as I am,still halfway pursuing truth, only reading can help approach supreme enlightenment."Learning is an endless process, the road ahead will be long and winding but our progress will be determined and promising.

Jinyu Bibliotheca (Mansion Knowledge) is located at the core area in Qianjiang New Town with a great number of upscale properties nearby;in order to extend the luxury positioningand highlight the theme of educational culture, the project aims to create a brand new high-quality premises whereinspiring family custom can shinethrough the contemporary roof.

The case issupposed to build a supporting sales department of Jinyu BibliothecaAdhering to the theme of the premises,combining with the international perspective and profound cultural background oftarget clients, the sales departmentis furnished in asupreme library-style,stands out as an unprecedentedbeacon not ever in the past or afterward, home or abroad, thanks to the inclusive and creative works conducted by the designer.

(Sales Department – Front Hall)
Adapted from the dome style of the world renowned British Library, along withcharming exotic bookshelf, the front hall

invites visitors in a stylish deluxe and elegant atmosphere. The design is dedicated to build a library as creative, resourceful and efficient as British Library, not only to reflect the external beauty but also to reward and present the innovative and resourceful connotation.

(Traditional Private School)
Upon accurate confirmation conducted by architects, a full size classic traditional private school magically brings San Wei Bookstore fromMr. Lu Xun's novel right in front of us, strikes us with the overwhelming charm of traditional culture.
Wise as you shall deserve such a place where you can embrace a bookstore, stay close with the changing world,but be aware it may subvert your outlook on life, shorten the distance between more brilliant people, and prepare you for a more glorious future.

The design style of the sales department isdedicated to quality. Instead of tangling in traditional interior architectural style, we integrate a collection of diversified elements in a free mix-matched eclectic way. Once you step inside get ready to be surprised.

Wandering through highly erected Europeanarchway passage provokes the feeling of entering the palace of knowledge and wisdom, receiving sacred baptism inglory and solemn ambience; the pattern of tile in the passage echoes the highly erected hollowed-out metal hallway, a subtle play of light and shallow perfectly connects them andforms a visual road towards heaven.The largerainy grey background wallrenders the simplicity and beauty of Chinese painting,collocating with antique tea set, painting & calligraphy and porcelain, etc.,exudes elegant Chinese ethnic chic.

The project is created to be a multi-functional environment, apart from traditional functions of sales department like cafe, display zone, negotiating zone, etc.;it allows to be operated as an unprecedented library.

What we are proud of is thedisplay zoneto which we appliedboth traditional andcontemporary techniques.Nestle into the comfortable sofa or relax yourself in the wood furnished leisure zone where providesfour precious articles ofwriting forpeople to demonstrate their talents.Sit down in the tea zone by the side and savor a cup of tea or enjoy the tranquil time. A nice display zone assures you a taste of life and an exclusive experience!

Adapted from the world renowned British Library, our unprecedented library is dedicated to refresh yourold impression, invite you to spend the best time reading and learning here and helpyou make the most sensible choice!

NOMINEE FOR BEST DESIGN AWARD OF EXHIBITION SPACE
最佳展示空间设计奖提名奖

浙江安道设计股份有限公司（中国杭州）

获奖项目/Winning Project

宁波镇海万科城滨湖体验中心设计
Vanke Lakeside Experience Center

设计说明/ Design Illustration

滨湖体验中心是镇海万科城项目的销售示范区，出于委托方对成本的控制，我们对滨湖体验中心进行了低成本改造，将成本控制在每平米300元以内。在极具现代特色的建筑空间里，融汇借景、框景等造景手法，体现一种区域文化的传承和贯通，呈现新潮与现代的空间氛围。项目在融合了水岸、绿植等自然元素的同时，还将工业化的集装箱元素融入其中，结合疏林草地、滨水休闲，将各个年龄层的用户体验引入景观空间，营造出具有艺术气息的互动性体验示范区。

As a demonstration area of Zhenhai Vanke Project, Lakeside Experience Center, due to the cost control of the Entrusting Party, conducted a transformation at low cost and cut the cost within RMB 300/square meter. In this exquisitely modernized architectural space, landscaping techniques such as borrowing and framing scenery are utilized to reflect the inheritance and connection of regional cultures, and help present an attractive contemporary spatial ambience. The project, integrating natural elements like waterfront & plants with industrial element like container, collocating scarce forest & grassland with waterfront leisure zone, along with the perfect application of sceneries catering for visitors of all ages, successfully creates an experience demonstration area full of art of aesthetics and interactivity.

IDEA-TOPS
艾特奖

NOMINEE FOR BEST DESIGN AWARD OF EXHIBITION SPACE
最佳展示空间设计奖提名奖

张立人
（中国台湾）

获奖项目/Winning Project

光景拟态
Lighting Scenery

设计说明/ Design Illustration

案件概要
这是一个不动产业的接待会所，针对业主所销售的案件在此做规划及洽谈，专门做为VIP客户的会谈用途。
开发意义：在都市丛林里刻意置入一个山形建筑，拥抱自
然与大地的意图十分彰显。

开发背景
群山绵延，乡野田间，传统屋瓦串连错落，高低层次穿插起伏于天地之间，这般使人愉悦的乡野闲情意象出现在位于台中七期的国泰河南路共销中心。这栋仅有一个楼层的建筑，被大都市钢筋泥墙围绕着，量体微小却散发着强大的自然力。常季设计师张立人以尊重自然、谦卑、谦让的态度，听到了大自然穿梭并驻足在都市中的美丽天籁，响应到被困于水泥牢笼的居住者们向往与林为邻的呼唤，再将设计意念转化为具象，保留近二分之一的基地给绿地重生。设计者透过设计让人与自然连结、互相感动，从而创造出的给都市居民的喘息空间。

设计理念
绿地中央为随意拉折线条的建筑体，由外观延伸到室内转折变换，但仅以很单纯的落地窗、天然材质构成，似在说着"我也属于这块绿地的一部分"试图隐身其中。为避开阳光西晒，设计师以拉折造型墙为入口，将建筑体的大面落地窗安排于南、北方，使进入空间的人，能转身间眼见为绿林，坐卧间触及接近绿地，大自然，在每一个角落都不曾缺席。

张立人某日清晨回到基地观察，看到一位父亲带着两个女儿上学，原是一脸疲惫，却在经过此基地时表情转变为欣喜，走在前方的女孩小手指着绿树，兴奋的回头对父亲描述所见，后方手牵手的父女也感染了相同情绪，三人满心喜悦，步伐更轻盈地继续走向目的地。设计师瞬间纪录下这幕，因这片绿地而引发人流露出对自然最原始的渴求、最真实的快乐，无需言语，设计不哗众取宠也能巧妙细腻的连结自然与人内心最深层的生活感动，同时也感动设计者本身。

创新点
以山形意象为概念中心、刻意退缩将屋外空地变成绿地，在寸土寸金的城市留给居民拥抱自然的一个机会。
设计师感言：空间设计尺度的掌握和一般商业产品不同，业主的要求也远比商业产品的要求更加繁琐复杂，从大尺度的规划到细部的拿捏，非常耗费心力，即便这样，当自己的创意落实至成品，是最令人欣慰的时刻。

Summary
This reception center belongs to a realty company, is for sales meetings, case arrangement and VIP customize service purposes.

Significance
This mountain-shape building is like being installed in an urban jungle, totally interprets the concept of embracing the nature and landscape.

Background
The reception center with a countryside style and surrounded by the mountains, but it located right in Tai-Chung city downtown, this deceiving spot at latest redevelopment zone of the city. Being surrounded by tons of concrete buildings, this tiny one-floor building is powerful and very much alive. Li-Ren, Chang, the designer, respects the mother nature with humble and modest attitude and has transformed this concept of " Metropolitan crave the neighborhood of nature" into something real by reserve half space for the greenery. Once again, inspires people with totally relaxation .

Theory
The building located right in the middle of a green area. The French windows and natural texture is telling us the object is part of the greenery. To avoid the western exposure, the designer made a folding wall in the entryway, and 2 French windows are facing the north and south, for people inside can appreciate the greenery view outside. Being closed to the nature is never this easy.
Li-Ren, Chang went to the construction site one day and happened to see a father taking 2 daughters to school. He looked exhausted, and when he walked by and saw this construction site, he suddenly become a different person. One daughter was ahead of them, stopped them with excitement and told them how beautiful the greenery is. They two felt the same way immediately and left there with joy and smile in their faces. He captured the very moment and all his hard work pays off. This enormous grass makes us desire for nature, that's the basic and fundamental need we have, no words or actions necessary, gets to the bottom of our hearts, and that inspires and touches the designer himself.

Innovations
Base on the mountain image concept, deliberately hold back exterior open space should be turned into green space, also left land-scarce city residents a chance to embrace nature.

When it comes to interior design, it's so different from any other kind of commercial products, and the demand from the clients could go extremely complicated. From the big picture to small details, it takes a lot of hard work and even so, when the job gets done, everything pays off, totally overwhelming.

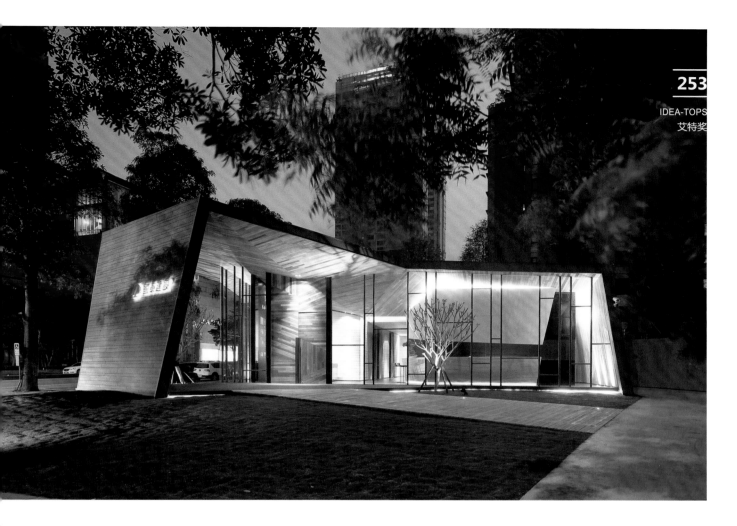

NOMINEE FOR BEST DESIGN AWARD OF EXHIBITION SPACE
最佳展示空间设计奖提名奖

陈恩育（中国杭州）

获奖项目/Winning Project

中交香滨国际售展中心
Champagne International Sales Center

设计说明/ Design Illustration

本案设于香滨国际配套商业建筑之内。入口位于建筑二层，通过一个长长的景观阶道引入。空间是有两个5.8m高的大方盒和一个宽敞连廊构成，模型大厅和洽谈大厅各占据一个盒子的大部分空间。设计师希望从售楼部惯常的富丽堂皇的营造模式中跳离出来，用干净利落的现代建筑语言和块面分明的雕塑手法来构建室内空间。并用温润的材质、色调来柔化过于硬朗的形态，以期望来宾在这宽敞高大的空间中能感受到温馨的、宾至如归的氛围。

模型大厅南与东向均是玻璃幕墙，具有极佳的视野和采光。内侧一面高5m，宽15m的橡木板饰面墙立体折转展开，一直延展到顶棚整面，从墙面折转空隙中发出的光带也顺着墙面顶棚放射出去，像是要与外面的天光接驳似的。西侧的显示屏墙面则是做了方型大框架，底面用外墙的建筑劈开砖，框体则是白色石材。起到一个收敛的作用，稳定了空间。

入口接待区紧邻模型大厅，与景观通道入口间增加了一个门斗做缓冲，经过二道门，一侧多边切面的大理石接待长台个性十足，而顶棚放射形的菱角线条使灯光显得丰富迷离。正对着的形象墙上发光的图形，正是项目标识中阳光、植物、飞鸟、人等元素的演化。人与自然融洽、优雅的生活，正是项目开发商和业主共同的愿景。

当宾客穿过长廊来到洽谈大厅，首先，面对的是一架通体洁白的巨大楼梯，座落在镶有一方静水的白色大理石平台上。楼梯面宽2.15m，长18m。这架增设的楼梯通往楼顶上的示范展示区。楼梯为钢结构，用装饰有竖向纹的片墙支撑，由于建筑楼板横梁的荷载不足，又增加了钢梁与立柱的连接满足结构条件，扶手栏板有二翼反向插向顶棚，而楼梯的顶端则是玻璃顶，白天强烈的天光洒下来，仿佛是天上放下的弦梯。其实这二翼对这个楼梯还起到了稳定平衡的作用。楼梯栏板的外侧采用了不规则菱角切面的造型，使楼梯更加生动，更富雕塑感。整架楼梯顺势而发，顺力而作。纯白色的楼梯在整墙木饰面的衬托下成为了整个洽谈区空间的视觉核心。

值得一提的是这里的水吧台，因为空间很大，设计师希望有一个立体的东西来撑起整个气场，又能和楼梯相对应，做一个视觉上的呼应或者是补位，就是一个4mx4mx4m的立方体框架，水平面和立面均扭转了一点角度。框内的吧台有一侧用内打光的水平的玻璃做，感觉是一个倾斜的装了水的大容器，容器是斜的，水是平的。

演奏台设计的比较轻松，就像一个小孩折了张小纸片，吹了口气，放大放大又轻轻落在厅里，听演奏的人便都聚过来了。

The case is set as a supporting commercial facility of Champagne International. The entrance is located on the second floor from whichyou can access the building through a long scenery corridor. The space is composed by two 5.8 meters high square boxes and a wide corridor, where model hall and negotiating hall occupy the majority space of each box.

In order to extricate from the routine luxurypattern of buildingsales department, I utilized clean and tidy contemporary architectural languageand well-defined sculpture techniques to allocate interior space, took advantage of mild texture and color to mitigate the hale state so as to create a warm welcoming chic in this spacious lofty building.

Glass screen walls were installedon the south and east sidesof the model hall to assure excellent scenery sight and lighting.Afive-meter high 15-meter wide oak wood decorated wall, which is folding open and extends to cover the entire ceilingwas implantedinside the building, the light belt reflected from the gap between turning walls projects light along the ceiling to invade the sky.The screen wall on the west was settled in a big square frame, the bottom used bricks withdrawn from the original external wall, the body of the frame used white stoneto echo the entire effect and stabilize the space.

The reception area at the entrance is adjacent to the model hall, so I planted a foyer in front of the scenery corridor. After accessing the second door, a long multilateral marble reception exudes strong characteristics, while the projective water chestnut lines craft a mysterious play of lights from the ceiling.The image wall standing opposing exhibits sparkling objects that represent for the evolution of sunshine, plants, birds, human, etc. Human and naturecoexisting in harmony is what the developer and proprietor wish to see.

Stepping into the negotiating hall, what welcomes the visitors is a huge white stair set above a tranquil white marble platform. The stair is 2.15 meters wide 18 meters long, leading towards the demonstration exhibition zone on the top floor.The steel structure of the stair is supported by a wall decorated with vertical stripes on the surface, due to the insufficient load of floor beam, I addedconnections between steel beam and column so as to meet structural condition, the handrail generated two wings reversely inserted to the ceiling, where sunlight shines through the glass ceiling and makes the stair appearas if a ladder putting down from the heaven. Actually, this pair of wings can alsostabilize the stair. The irregular water chestnut sections that were applied to the outside of stair bestowed it with avivid image that seemed more like a sculpture. The construction of this stair takes advantage of and gives priority to all available resources. When viewed against the wooden surface wall,the pure white stair becomes the core of the entire negotiating area.

What is worth mentioning is that the water platform, considering the mass space around, has to be dynamic and remain contrasting to the stair, so I created a 4*4*4 meter cubic frame, twisted horizontal and vertical angles so as to bring forth visual echo with the stair. Set a horizontal glass base inside the platform to project light from one sidelike a slant huge containerfilled with water inside.

The performance stage adopts a leisure design, it seems like a kid folded a piece of paper and blew a breath of air to it, then it gradually grew bigger and bigger, finally landed in the hall and attracted crowds of people to see the performance.

259
IDEA-TOPS
艾特奖

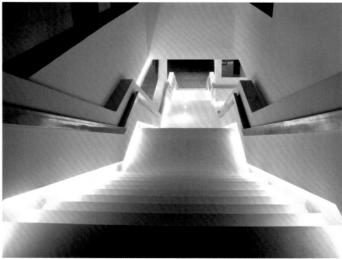

NOMINEE FOR BEST DESIGN AWARD OF EXHIBITION SPACE
最佳展示空间设计奖提名奖

钟荣海（中国厦门）

获奖项目/Winning Project

组合
Combination

设计说明/ Design Illustration

这是位于德国杜塞尔多夫的一次零售业展览，三年举办一次。HUMKA作为国内模块化货架开发的领先企业之一参加了此次展览。展览一般都是短期的行为，三到五天。展会结束后，所有的物料都必须拆除，大部分都是当作垃圾扔掉。每次展会结束后整个展馆就是一个大型垃圾场，但在国外的展会，这种情况会好很多，因为处理垃圾的费用相当高昂，根据不同的种类大概一个立方要100~300欧元左右，所以绝大多数的展位都是组装设计，并且可以重复利用。HUMKA展位的面积是100m²，所有的物料光是运输费用就是不小的开支。因此，扁平化的包装、灵活的组合式搭建成了唯一的选择。展台的设计主要使用了半透明亚克力，利用其导光属性，现场仅使用了少量的灯光即达到了照度要求，同时采用模块化的组合方式，搭建快捷，只用了3~5个人工，三天就完成了搭建工作。

This is a retail industry exhibition located in Dusseldorf of Germany and held once every 3 years. HUMKA participates in the exhibition as a leading enterprise of modular shelf development in China.

The exhibition usually lasts for a short period from 3 to 5 days. After the exhibition, all the materials must be removed and most of them are discarded as wastes, so that the whole exhibition hall is a large waste yard after each exhibition. In foreign exhibitions, there are so many such cases. Since the expenses for waste treatment are very expenses (generally € 100–300 per cubic meter according to different type), most booths adopt assembly design and can be reused.

The booth of HUMKA is 100m2. The transportation expenses for all the materials are considerable. Therefore, flat packing and flexible combined construction is the only choice. The booth is designed to adopt translucent acrylic. With its light transmission property, only few lights are adopted on the site. Meanwhile, the modular combination mode is adopted, so that the construction is completed by 3-5 persons in 3 days.

What is worth mentioning is that the water platform, considering the mass space around, has to be dynamic and remain contrasting to the stair, so I created a 4*4*4 meter cubic frame, twisted horizontal and vertical angles so as to bring forth visual echo with the stair. Set a horizontal glass base inside the platform to project light from one sidelike a slant huge containerfilled with water inside.

The performance stage adopts a leisure design, it seems like a kid folded a piece of paper and blew a breath of air to it, then it gradually grew bigger and bigger, finally landed in the hall and attracted crowds of people to see the performance.

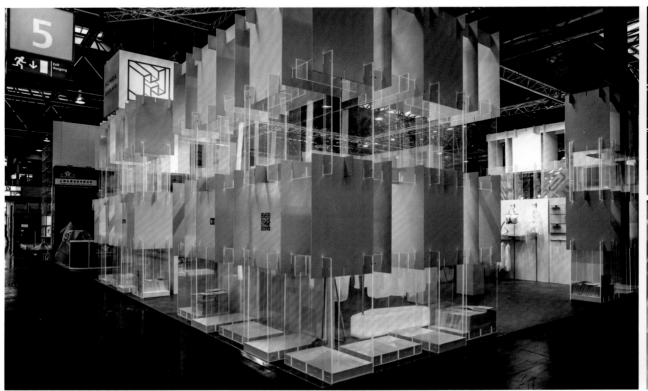

263
IDEA-TOPS
艾特奖

ATELIERCENTRAL

265
IDEA-TOPS
艾特奖

获奖者/ The Winners
Atelier Central（AC）
（葡萄牙）

获奖项目/Winning Project
Montemor-o-Velho学校/ Montemor-o-Velho School

获奖项目/Winning Project
Montemor-o-Velho学校
Montemor-o-Velho School

设计说明/ Design Illustration

校园里有一条用棕色混泥土建造的廊道，连接着东部和西部两个入口。廊道两旁，颜色各异的混凝土建筑物连接着不同的通道，使既有建筑和新建建筑区分开来。

聚集空间和非聚集空间有逻辑地连接着三个学院。

设计师对分散的教室和行政办公区域进行了改造，通过一个回廊、无障碍区、通道、会议区、生活区和操场将它们连接为整体。

校园内增添了新的设备，图书馆、食堂、覆盖区以及多功能大厅分散在各个区域，将广场、草地、庭院隔开——这要求设计师们谨慎地规划室内与周围环境。

该项工程的方案由Parque Escolar和学校共同确立。校园里有一座八十年代风格的展馆，但学校想要营造一种新的文化和学习氛围，这就像是当代的个体环境与空间环境之间的关系。

通过透明的环境，空间和不同活动区的聚合使校园变得开阔，有利于校园设备的共享，也增强了不同学院学生之间的交流。

该项目试图营造一种不折不扣的非正式的教育氛围。

Trace a structuring path through the School Campus, in brown apparent concrete, that unites the two entrances at East and West, along which, marked by connecting / access tunnels in concrete pigmented in different colours, the existing and new buildings are distributed.
A gathering and ungathering space assumed in a logic of uniting the three schools.
Remodel the disperse Classroom and Administrative Support Blocks, uniting them through a cloister, accessibilities area, circulation, meeting, living and covered playground.
Conceive the new equipment, Library, Dining hall, Covered Field and Multipurpose Hall, as isolated objects that appoint squares, mark meadows, reveal courtyards and timidly project the interior over the surroundings.
The approach to the program, defined by the Parque Escolar and the School's Administration, results from the confrontation of the pavilion like school of the eighties versus new and desired learning culture thought in accordance with contemporary relations between individual / spatial surroundings.
Open the school to the community stimulate sharing of equipment, of experiences between students of several Schools through the transparency between environments, the spatial confluence and the convergence between different activities.
Try to stimulate and provide an uncompromised and informal schooling.

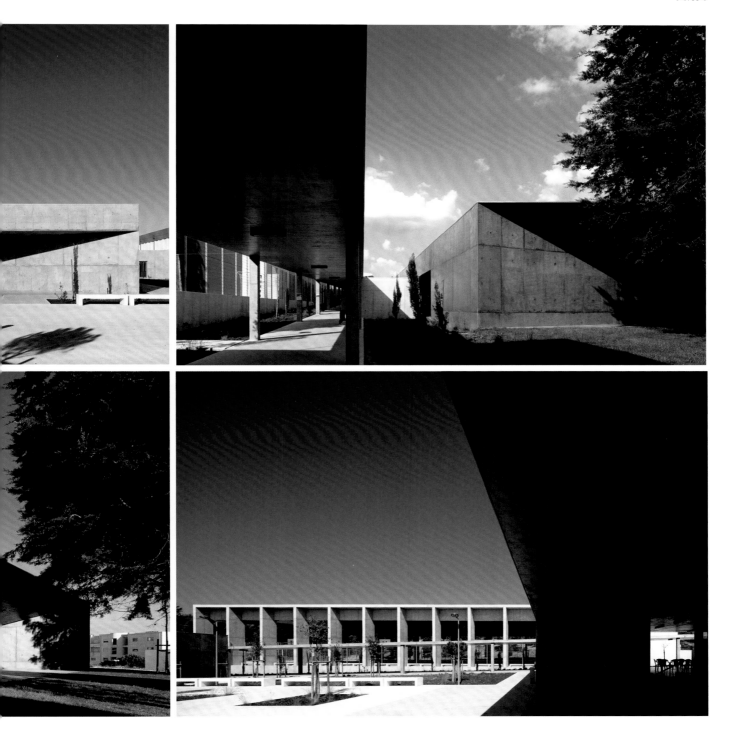

270
IDEA-TOPS
艾特奖

获奖评语

获奖作品创造了一个有趣而又迷人的文化空间，通过清晰的角度和光线使建筑物变得开阔，促进人们交流和开展探索活动。

The winning entry crafted an engaging and alluring cultural space, opening up the building through clear angles and light, encouraging engagement & exploration.

NOMINEE FOR BEST DESIGN AWARD OF CULTURAL SPACE
最佳文化空间设计奖提名奖

ATELIERCENTRAL

Atelier Central（AC）
（葡萄牙）

获奖项目/winning Project

Bicesse幼儿园
Kindergarten in Bicesse

设计说明/ Design Illustration

该项目的设计理念是在孩子们充满想象力的世界中寻找参照物，并将它们融入到空间之中。整个建筑拥有统一的常规功能区域，采用透明和不透明的材料以及圆柱形的天窗。许多物品都来自于"乐高积木"，它们通过不同的形式组合，分布在走廊上，展现孩子们在幼儿园里的成长历程。走廊区域设有具有保护性的托儿所，入口旁还为5岁以上的孩子设计了公园。走廊作为该建筑的轴线，连接着儿童房、管理区、服务区。使用半透明玻璃材料建造的走廊，白天为室内带来光线，夜晚则成为一个巨大的"盒子"，照亮外部空间。环绕的墙壁形成了两个休闲区域——露台和娱乐区。露台位于食堂旁边，有一棵树和一面水镜，相对比较封闭，营造安静的氛围，使人放松。娱乐区朝南，临近儿童房，在这里可以看到美丽的海景。

The idea was to look for important references of the children's imaginary world. The building is constituted by pure and regular volumes, either opaque or transparent, with some cylindrical skylight cutting the ceiling. They are parts of Lego, grouped and disposed throughout a longitudinal corridor of distribution that represents children's evolution in the kindergarten. They start in the nursery, in the more protected area, and finish, when they have five years old, in 2 o Park, next to the entrance or exit. Being the structural axle of the building, this corridor is the

frontier between the side of the children's rooms and the other of the administrative and service areas. Its construction in translucent glass accents this idea because it becomes the main source of light for the interior, during the day, and it will be a huge box illuminating the exterior, during the night. A surrounding wall confines two areas of leisure: A patio, more contained, relaxant and tranquilization, next to the refectory, with a sobreiro and a mirror of water. A recreation, for where if they guide and they open the rooms of the children.

NOMINEE FOR BEST DESIGN AWARD OF CULTURAL SPACE
最佳文化空间设计奖提名奖

Juergen Mayer H（德国）

获奖项目/Winning Project

KA300纪念馆
Jubilee Pavilion KA300

设计说明/ Design Illustration

KA300展馆于2015年6月20日正式开放，其位于德国卡尔斯鲁厄宫廷花园，用于城市Jubilee的展馆揭幕，这是由J．MAYER H与其合作伙伴建筑师事务所共同设计的。为了庆祝卡尔斯鲁厄这座城市成立300周年这一事件，该展馆建成于该城市的宫廷花园。在夏天节日期间，各种音乐会、戏剧表演、阅读、电影放映以及展览会将在开放的结构里举行。该展馆提供了一个大型的礼堂并有一个舞台：它是该城市周边举行欢乐活动的中心，也是拥有咖啡馆的会议地点的中心。该展馆的扭曲的模式是参照巴洛克规划的卡尔斯鲁厄城市的严格的几何、径向布局，以宫殿作为焦点，将其转化成多种线条的空间领域。在该结构的多个楼层上，展览平台、休息空间和观景平台出现了。在与投标竞争时，J．MAYER H和其合作伙伴建筑师事务所以及Rubner Holzbau收到了委员会对于他们提出的使用条型钢进行的动态木建筑。展馆于2015年2月开始施工。所有木材由Rubner提供，已准备好用于现场安装，以便保证在宫廷花园的地面上进行一个较短、及时的施工。该周年庆祝活动从2015年6月17日至9月，活动结束后，临时的展馆将被拆除，随后运输至另一个位置，并得到再利用。

Pavilion for the City Jubilee, Schlossgarten, Karlsruhe June 20th, 2015 marked the opening of the KA300 Pavilion in Karlsruhe, designed by J. MAYER H. und Partner, Architekten. To celebrate the three-hundred year anniversary of the founding of the city of Karlsruhe, this temporary event pavilion was erected in the city's Schlossgarten. During the festival summer, various concerts, theatre performances, readings, film screenings, and exhibitions will be held in the open structure. The pavilion offers a large auditorium with a stage: it is the center of the jubilee activities around the city and a meeting point with a café. The twisted pattern of the pavilion refers to the strictly geometric, radial layout of the Baroque planned city of Karlsruhe with the palace as the focal point, transforming it into a spatial field of lines. On several layers in and on the structure, exhibition platforms, resting spaces, and viewing platforms emerge. In a competition with tendering, J. MAYER H. und Partner, Architekten and Rubner Holzbau received the commission for their proposed dynamic wood construction using bar profiles. Construction started in February 2015. All wood provided by Rubner is supplied ready for installation on the site in order to guarantee a short and timely construction on Schlossgarten grounds. After the closing of the anniversary celebration from June 17th to September 2015, the temporary pavilion will be dismantled, transported and reusd afterwards in another location.

NOMINEE FOR BEST DESIGN AWARD OF CULTURAL SPACE
最佳文化空间设计奖提名奖

康悦（中国青岛）

获奖项目/Winning Project

离大海最近的公立美术博物馆
（烟台美术博物馆）
The Seashore Musume

设计说明/ Design Illustration

设计主旨：吐故纳新融合之美。

设计背景：具有地域性特质的渔家文化和海洋文化，深深影响着烟台的艺术文化，伴随着新文化、新艺术长足发展，当地的传统艺术文化得以与之细密的交汇。

主案构思：依托设计主旨，分析本案的外观，同时考量其功能特质，内部空间主要使用现代的设计手法，运用洗练的线条和秩序感的构成，让材料质朴的表达自身，营造出简美的艺术空间。

空间构成：一层采用辐射式和串联式，负一层采用厅式进行动线组织和空间设置，增强空间的交互性。

平面材料：半抛白色花岗岩、黑色页岩、白色石英文化石、樟子松实木、灰泥、方钢等。

Theme of Design: exhale the old and inhale the new, the beauty of fusion
Design Background: as regional characteristics, fisherculture and ocean culture have deeply influenced the formation and development of art & culture in Yantai, due to thesubstantialdevelopment of new culture and art, the local traditional culture & art integrate and interact closely with the new trend.
Project Conception: taking into consideration of the design theme, the appearance and the property the museum, the design of interior space mainly utilizes modern techniques, combines with succinct lines and ordered construction, allows the materials to express themselves, thus creates a simple beauty of art space.
Space Construction: the first floor applies radiant-style and tandem-style; the negative first floor applies hall-style, using circulation and spatial allocation to enhance the interactivity of space.
Planar Materials: semi-polished white granite, black shale, white quartz culture stone, solid Pinus sylvestris, plaster, square steel, etc.

284
Best Design Award
Of Commercial Space
最佳商业空间设计奖

艾特奖
最佳商业空间设计奖
BEST DESIGN AWARD
OF COMMERCIAL SPACE

INTERNATIONAL SPACE DESIGN AWARD

获奖者/ The Winners
朱志康
（中国深圳）

获奖项目/Winning Project
成都方所书店/ Fangsuo Bookstore

获奖项目/Winning Project

成都方所书店
Fangsuo Bookstore

设计说明/ Design Illustration

早在千年前中国人就为了寻找古老智慧而不辞劳苦，获取经书，大慈寺就是唐代玄奘前往西天取经前修行的地方。经书和书店都是智慧的宝藏，因此有了藏经阁的概念,并且应该如圣殿一般的庄重。

书中收纳了古今中外的历史和智慧，根植于人类已知的世界，求索未来。设计师在整个空间里运用了星球运行图、星座元素来增加浩瀚的宇宙视野。台湾设计师朱志康在空间设计上面运用了很多高压后释放的设计手法，让人体会通过神秘隧道进入圣殿的感动！

顺应四川人闲适喜爱交流的"窝"和"摆"的生活态度，书店中随处都有可以看书的小空间和相互交流的场所。

Chinese people sought for ancient wisdom and scripturespainstakingly a thousand years ago. Daci Temple is the place where Xuanzang of the Tang Dynastymeditated and practiced Buddhism before he embarked on a pilgrimage to the west for Buddhist sutras. Scriptures and bookstores are treasures of wisdom; hence we have the concept of depositary of Buddhist texts which should be solemn as a grave temple.

Books are the carriers of the history and wisdom of all times and all countries, rooted in the known world of human beings for search for the future. Therefore, star running chart and constellation element are used to enhance the vision of the vast universe in the whole space. Zhu Zhikang, a designer from Taiwan, applies multiple techniques released after high pressure in space design, making people taste the exciting feeling of passing through the mysterious tunnel to enter the grave temple!

Regarding spatial layout, he followed the concepts of "nest" and "gossip" ideas deeply rooted in Sichuan people's attitude towards life. Correspondingly, the bookstore offers plenty of small spaces to read as well as coffee tables to allow its visitors to communicate and enjoy.

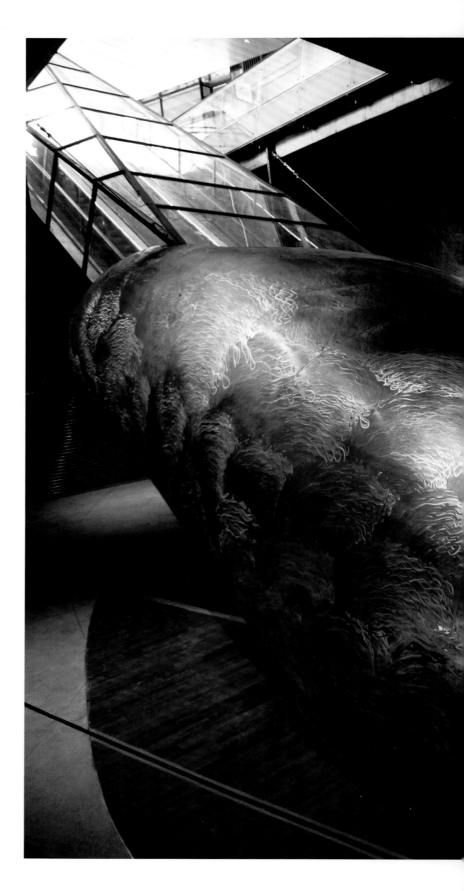

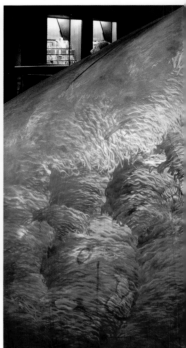

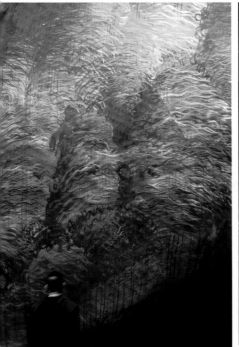

获奖评语

这个很有创意的商业空间邀请客人来探索一个概念商店的令人惊奇的地下世界。
This very creative commercial space invites the visitors to discover the amazing underground universe of a concept store.

NOMINEE FOR BEST DESIGN AWARD OF COMMERCIAL SPACE
最佳商业空间设计奖提名奖

Atelier Central
（葡萄牙）

获奖项目/Winning Project

BMWi里斯本销售展厅
City Sales Outlet BMWi Lisbon

设计说明/ Design Illustration

该项目的设计理念是让汽车展览空间带上具有神秘色彩的动感白色"面纱"，从而使其与周边环境截然不同。

透明和半透明的玻璃表面让光线交错折射，形态各异的光影、图像、影片消除了整个空间的边界感。

自发辐射的光线围绕着空间不断变幻，让各个角落和边界都消失不见。

一方面，一盏垂直的蓝色LED灯照射在有着均匀光线的墙壁上。

另一方面，圆形的车头灯忽明忽暗。

天花板上的灯光随机分布，将我们带入一个高速运动的世界里。

在天花板较高的区域，高色度的日光照射在展厅区，在天花板较低的区域则运用了温暖的光线，欢迎着人们来到接待区域。

整个设计营造了一种静谧的氛围，使得展示的汽车能够表达出该项目本质的重要性。

卫生间、储藏室、技术区的门，以及通向技术区的通道都被隐藏在精心设计的玻璃表面和假天花板下的活板门后面。

展厅内还有一个为BMW项目量身定做的汽车展示平台！标准严格的可持续性和尊重环境是宝马（BMW）集团关心的主要问题，因此这个项目也体现了这两点。

整个空间在LED灯的照射下光芒四射。设计师更换了展示窗口的玻璃，从而降低了太阳辐射，并减少热透射率。此外，空调设备中还安装了热回收系统。

Designing a mysterious dynamic white veil that segregates the cars on display from the surroundings.
An alternating transparent and translucent glass surface that stimulates reflections, reveals/conceals different light shapes, discloses images, frames videos, and eludes the perception of the boundaries.
The unending movement of the spontaneous line surrounds the available space, hiding corners and eliminating boundaries.
On one side, a wall of uniform light ending with a vertical blue LED light that refers to.
On the other side, round car headlights that switch to become visible, and then switch off, and disappear.
The lighting in the ceiling, apparently scattered at random, takes us to the world of high-speed movement.
Day light, with high chromatic definition, over the showroom area, with its higher ceiling, and warm light welcoming people to the reception areas, with lower ceiling height.
A tranquil design that allows the cars on display to express the importance of the project essence.
The doors to the bathroom, the storeroom, the technical areas and the access to the latter disappear, concealed by the deliberate detailing of the glass surface and the trapdoors in the false ceiling.
Hey presto, and we have a car display platform as defined by the BMWiProgramme!
Strict criteria of sustainability and respect for the environment are major concerns of the BMW Group and are reflected in the project.
The space is lit by LED lights. The glass in the display window was replaced to gain solar radiation control and reduce the coefficient of thermal transmittance. A heat recovery system was coupled to the air-conditioning equipment.

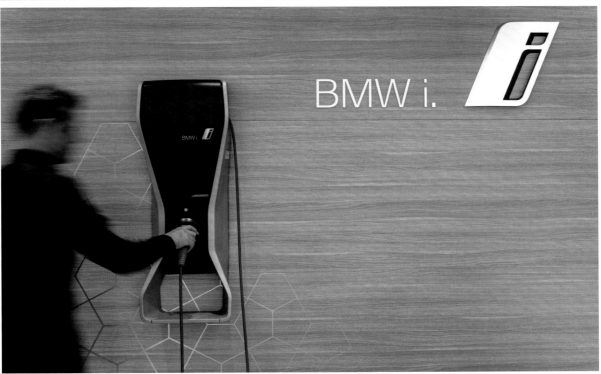

NOMINEE FOR BEST DESIGN AWARD OF COMMERCIAL SPACE
最佳商业空间设计奖提名奖

杨胤（中国沈阳）

获奖项目/Winning Project

華根卫浴商展空间
Huagen Bathroom Exhibition Space

设计说明/ Design Illustration

华根卫浴商展空间是一个卫浴产品的展示厅，位于红星美凯龙的一层。展区面积非常小，仅为14.2m×7.4m，平面规整、方正。
这是一个可以自由穿行的家具装置。设计师在规则的空间内引入45°的斜线，在矩形的体积内切割出若干个折线型的盒子，这些动感的盒子经过二次切割，形成高度集成化的家具装置。不仅几乎所有的卫浴展品、配件均被包含在这些家具之中，而且还容纳了接待、会议、酒吧等多种空间。盒子之间的缝隙则是连续的走道，人们可以在这些折动的家具装置之间自由穿行。
这里是一个小型的展厅，同时也是一个巨大的展示装置。它是内敛的，也是开放的，鼓励人们去体验行走在家具内的愉悦。

材料
墙面：美国橡木实木强化地板。
天花：灰色铝百叶、发光软膜。
地面：美国橡木实木强化地板、黑色哑光地面砖。

Wagenfeld Bathroom Trade Show Space is an exhibition hall of bathroom products located at the first floor of Red Star Macalline. The exhibition area is very small, only covering 14.2m*7.4m. The plane is neat, upright and foursquare.

This is a furniture device that people can pass through freely. We introduce a 45° oblique line in the planned space and cut out several fold-line boxes in the rectangular volume. Such dynamic boxes form a highly-integrated furniture device via secondary cutting. Nearly all bathroom exits and accessories are concluded in these futures. Moreover, multiple spaces for reception, meeting and bar are contained. The gaps between the boxes are continuous walkways that people can pass though freely.

This is a small exhibition hall, but it is a huge display device. Being restrained and open, it encourages people to experience the joy of walking in furniture.

Materials:
Wall surface: American oak solid wood laminate flooring
Ceiling: Gray aluminum shutter and luminous soft film
Floor: American oak solid wood laminate flooring and black matte floor tile

NOMINEE FOR BEST DESIGN AWARD OF COMMERCIAL SPACE
最佳商业空间设计奖提名奖

邱春瑞（中国深圳）

获奖项目/winning Project

莲邦广场艺术中心
Lianbang Square Art Center

设计说明/ Design Illustration

用地位于珠海横琴特区横琴岛北角，紧邻十字门商务区，东北面紧邻出海口，享有一线海景，景观资源丰富；与澳门一海之隔，更可观澳门塔、美高梅、新葡京等澳门地标建筑；东面距氹仔经200m，地理位置优越。

整体项目从"绿色""生态""未来"着三个方向出发规划。从建筑规划设计阶段开始，通过对建筑的选址、布局、绿色节能等方面进行合理的规划设计，从而到达能耗低、能效高、污染少，最大程度的开发利用可再生资源，尽量减少不可再生资源的利用。与此同时，在建筑过程中更加注重建筑活动对环境的影响，利用新的建筑技术和建筑方法最大限度的挖掘建筑物自身的价值，从而达到人与自然和谐相处的目的。

建筑概念设计
整体建筑造型以"鱼"为创意，采用覆土式建筑形式，整个建筑与周边环境融为一体，外观像一条纵身跃起的鱼儿。该建筑与周边环境充分融合，覆土式建筑形式可供市民从斜坡步行至艺术中心顶部休闲娱乐，且同时可观赏到珠海、澳门景观。建筑中心区域通过通透屋顶的处理，建立室内外的灰空间，从视觉上形成室内外一体景观，做到了室内、室外的充分结合。

建筑周边结合园林绿化设计，通过水景过渡及雕塑、装置艺术品等的设置，增加艺术氛围，形成滨海的、艺术的、人文的、自然的公共休憩场合。

雨水回收：通过采集屋面雨水和地面雨水统一到达地面雨水收集中心，经过雨水过滤再利用输送给其他用途，如卫生间用水、景观用水和植被灌溉。

能源回收：建筑外墙体通过使用能够反射热量的低辐射玻璃，尽可能多的引进自然光，同时减少人造光源。建筑覆土式设计采用自然草坪，在一定程度上形成局域微气候，减少热岛效应、隔热保温，能够高效的促进室内外冷热空气的流动，降低室内温度到人体接受范围。

"室内是建筑的延伸"
首先考虑建筑外观以及建筑形态，在达到审美和功能性需求之后，把建筑的材料、造型语汇延伸到室内，并把自然光及风景引进室内，将室内各个楼层紧密联系，人文环境相互律动，是室内空间的节奏。

动线安排
室内部分共分为两层——展示区域和办公区域，客户在销售人员的带领下首先会经过一条长长的走廊，到达主要区域，在这里阶梯式分布着模型区域、开放式洽谈区域、水吧台以及半封闭式洽谈区域。在硕大的类似于窈窕淑女小蛮腰的透光薄膜造型下，这里可以纵观整个综合体项目的规划3D模型台。阶梯式布局采用左右对称设计，左边上、右边下，一路上都可以领略到窗外的风景。靠近澳门的这一面，采用全落地式低辐射玻璃，在满足光照的前提下，可以很好地领略澳门的风景。绕着一个全透明的类似于椎体的玻璃橱窗-这里也是整个不规则建筑体最高处，达到12m高度——可以到达2层区域的办公区域，在挑高层那一侧可以清楚的看见一层的主

Thesite is located at the northcorner of Hengqin Island in Zhuhai,withinclose proximity of Shizimen Central Business District.Itsnortheast is close to the sea and enjoys excellent sea view and landscape resources.The site isseparated with Macau only by a river and enjoys the view of the landmarks of Macau suchas Macau Tower,MGM and Grand Lisboa. With Taipa 200m to its north, the site enjoys an excellentlocation.

We plan the overall project from the principle of "green", "ecology" and "future".Starting from the period ofarchitectural planning,the designers arrange reasonably the site,layout and energyconservation of the building, so as to meet the demand of low powerconsumption,high efficiency and low level of pollution, as well as to exploit renewable resources at the greatest extent and minimize the use of non-renewableresources.Meanwhile, we pay more attention to the influence of the construction activity on environment and utilize new construction technologyand technique in the project to excavate the value of the building itselfat maximum,ending up with the consequence of harmonious relation between human andnature.

Conceptual architectural design
Inspiredphilosophy of the outline is from fish. The building is constructed in an earth-covering form.Hypostatic unionwith the surroundings makes it hard to distinguish the demarcation line.It appears like a leaping fish. Citizens can walk to the top of the art center for leisure and recreation from the slope and enjoy the landscapes of Zhuhai and Macau as well. Thecentral area of the building is made translucent to develop a grey spaceconnecting interior and exterior, forming an integrated landscape in visual sense.Via making the most use of surrounding landscape design including the installation ofsculptures and artworks, the artistic sense is enhanced. In this way, a public recreation site integrating with coast,art,humanity andnature is created.

Rainwater collection:Rainwater on the roof and ground is collected into the ground rainwater collection center. After filtration, the rainwater will be used for otherpurposes such as toilet,landscape and irrigation.

Energy conservation:The exposed walls are occupied with low-E glass capable of reflecting heat, importing natural lighting at most and reducing the useof artificial lighting. The earth-covering design with real grass brings out topicalmicroclimate ending up with the reduction of heat island effect andperseverance of

warmth.

"Interior is the extension of a building"
First of all, the designers should take the architectural appearance and form into consideration. After the aesthetic and functional needs are met, the designers should extend materials and design conception inheriting from the entire architecture to the interior. To invite natural lighting and views into the interior, they need to connect all floors intensively. The interchange between human and environment acts as rhythm of interior space.

Flow lie arrangement
The interior covers two floors—display area and office area. When a customer is led by a salesman to such main areas as model area, open negotiation area, water bar counter and semi-closed negotiation area, a long corridor must be passed by first. With the transmitting film modeling, the planned 3D model platform of the overall project can be viewed here. The step layout adopts bilateral symmetry design, with the left for going upstairs and the right for going downstairs. The scenery outside the window can be viewed all the way. Floor-standing low-E glass is adopted for the side close to Macau, so that the scenery of Macau can be well appreciated on the premise of good sunlight. Walking upward around the transparent glass window to highest height of 12m, you will get at the office area on the second floor where the main working area can be seen. To have a panoramic view of the landscapes of Macau and Hengqin, you need to go to the roof of the building by the arc-shaped stairs inner the cylindrical vitreous body.

NOMINEE FOR BEST DESIGN AWARD OF COMMERCIAL SPACE
最佳商业空间设计奖提名奖

Ramon Esteve(西班牙)

获奖项目/Winning Project
沙波时尚店
Chapeau Fashion Store

设计说明/ Design Illustration

该多品牌商店是由建筑师Ramo nEsteve设计的。在商店里，我们可以在一个800m²的地面上发现男人和女人时装。

Chapeau时装店拥有25年的经验，现已放置了陈列柜，以展示奢侈品以及在西班牙的时尚潮流。

这个新的空间位于瓦伦西亚Herna nCorte s大街5号。男人和女人时装店共存于一个建筑空间里，设计了一组反射镜和棱镜，其中各式服装都是主角。沿着它的长度，被白镜更衣室的一条线分开，该空间在一个巨大的天窗上达到顶点，该天窗随着一个黑色玻璃墙上升，使得该旅行成为一种真实的视觉体验。就此事而言，诸如普拉达、Marn、古奇、斯特拉•麦卡特尼、亚历山大•王、巴尔曼、圣罗兰、朗万和MiuMiu等专用于妇女的时装品牌，放置在该商店的右翼，虽然其他品牌比如专用于男性的时装品牌
汤姆•福特、桑姆•布郎尼、普拉达、蒙克莱、巴尔曼、
尼尔•巴瑞特或外来的品牌，则放置在该商店的左翼。

考虑在这个新的空间里，Chapeau时装店集团共有三个商店，位于该城市最繁忙的商业环境里。在同一条街上，Chapeau时装店第3家分店（位于瓦伦西亚Herna nCorte s大街16号）专门经营年轻人时装，而Chapeau鞋店（位于Cirilo Amoros大街39号）用于经营女性配饰。其所有商店基于卓越的哲学理念，他们取得成功的关键一直是专业选择各式衣服。自从业主皮拉尔•Puchades和何塞•Tamarit在1987年推出了第一家店铺，Chapeau时装店就有了巴黎、米兰和纽约最好的收藏品展厅。其制定规则的能力在大量来自整个西班牙的客户中建立了忠诚度。"我们的客户是一种衣着简朴而优雅的女人，而不是一种季节性的女人。"何塞•Tamari解释说。
现如今，Chapeau时装店是西班牙顶级的多品牌商店之一。"所有商店必须改造，不仅是其产品，而且包括空间。"

空间建筑的一个大型黑色钢选框通过其开口予以照亮，这是该商店玻璃幕墙的框架，展示了分别用于男性和女性的入口，作为显示其展销品的场景。该商店窗口通常是以摆放人体模型的空间为特征，在本例中，其打破自然障碍和视觉界限，用整个商店作为背景。
在商店内，从过道开始，会通过一系列场景，其主体是两个大型LED屏幕以及由哈维尔•Santaella为Chapeau时装店制作的录像艺术作品。下一个场景是来自于墙壁的长片，在此你会发现来自于地面的石材计数器，这里是连接男性和女性区域之间的空间。到达更衣室后，这些区域在视觉上又有区别了。在末端，该空间用一个高背景的天窗被再次连接起来，使得该商店内部充满自然光。
随着空间的进展，该商店在一些区域着重显示了各种展销品的身份，以便给每个展示的品牌以适当的评价。

这个项目是由纯粹的几何定义的，由三个基本类别表示。一方面，各套棱镜的实现是通过各种物品比如该中心里的反射镜盒子。这些棱镜被用于更衣室。另一方面，白色光线在周边聚集，大型垂直黑镜子在更高的空间。

下一个类别是有关来自于天花板和天窗的各种平面。沿纵向路径的一连串的斜平面可以容纳各类设施而不损失高度。这些平面意味着到达该商店的末端，在此它们转变成斜角玻璃片材并关闭水平面上的天窗。

最基本的类别对应于衣架以及衣服存放和悬挂所用精美家具而形成的线条。这个元素也见于开口处，在整个空间、破裂平面以及生成的边缘都随处可见。

The new reference multibrand store is designed by the architect Ramo nEsteve. At the store we can find man and woman fashion in a 800 square meter surface.

With 25 years of experience, Chapeau has placed as a showcase for luxury and fashion trends in Spain.

This new space is located in 5, Herna nCorte s Street of Valencia. Men and women fashion coexist in an architectural space designed with a set of mirrors and prisms in which the garments are the leading character. Divided along its lenght by a line of white mirror changing rooms, the space culminates in a huge skylight that rises over a black glass wall making the journey an authentic visual experience. In this matter brands like Prada, Marni, Gucci, Stella McCartney, Alexander Wang, Balmain, Saint Laurent, Lanvin and MiuMiu for women are placed the right wing of the store, while other brands like Tom Ford, Thom Browne, Prada, Moncler, Balmain, Neil Barret or Band of Outsiders are arranged at the left wing dedicated to men.

Taking on account this new space, Chapeau group has a total of three stores located in the busiest commercial environment of the city. In the same street, Chapeau 3 (16, Herna nCorte s Street) is dedicated to a younger fashion and Chapeau Shoe in (39, Cirilo Amoros Street for women accessories.

Based on the philosophy of excellence in all its stores, the key for their success has always been the expert choice of clothes. Since the owners Pilar Puchades and Jose Tamarit launched the first shop in 1987 Chapeau has had the best collections of the showrooms of Paris, Milan and New York. Its capacity for prescribing has built loyalty in a significant number of clients from

all over Spain. "Our client is an elegant woman who dresses simply. Not a woman of a season," explains Jose Tamarit.

Today Chapeau is among one of the top multibrand stores of Spain. "Stores have to renovate, not only the product but also the spaces."

ARCHITECTURE OF SPACE

A large black steel marquee is illuminated through openings, it is the frame for the glass facade of the store, presenting separate entrances for men and women as scenes for showing the collections. The shop winfdow is usually characterized by the space where mannequins are, in this case is breaking natural barriers and visual boundaries having the entire shop as background.

Inside the store you pass through a sequence of scenes beginning with the access, dominated by two large LED screens with video art compositions produced by Javier Santaella for Chapeau. The next scene is defined by long pieces that emerge from the walls where you find the stone voulme counters emerging from the ground connecting spaces between male and female areas. Reaching the changing rooms the areas are visually separate again. At the end the space is joined again in a high background topped by a skylight that floods with natural light the interior of the store.

This progression of spaces organizes the store in areas that highlight the identity of the collections in order to give appropriate value to each brand exposed.

The project is defined by pure geometry which is represented by three basiccategories. On one hand, sets of prisms materialized through various items such as boxes of reflecting mirrors in the center. This prims are used as changing rooms. In the other hand white light volumes on the perimeter and large vertical black mirror at the higher space. The next category is about the planes that form the ceilings and skylight. A succession of inclined planes along a longitudinal path can accommodate facilities without losing height. These planes lead to the end of the shop, where they transform into angled glass sheets that close the skylight on a horizontal plane.

The most basic category corresponds to the lines formed by the clothes hangers and the fine pieces of furniture where clothes rest and hang. This element is also found in the openings that invade the whole space, breaking planes and generating edges.

The formal language defined by the basic geometries of solid, plane and line,materializes with glass, mirrors, stones and metals in white, gray and black tones to achieve the ideal atmosphere for exhibiting the highest quality product in an attractive play of reflections, shades and lights.

What is worth mentioning is that the water platform, considering the mass space around, has to be dynamic and remain contrasting to the stair, so I created a 4*4*4 meter cubic frame, twisted horizontal and vertical angles so as to bring forth visual echo with the stair. Set a horizontal glass base inside the platform to project light from one sidelike a slant huge containerfilled with water inside.

The performance stage adopts a leisure design, it seems like a kid folded a piece of paper and blew a breath of air to it, then it gradually grew bigger and bigger, finally landed in the hall and attracted crowds of people to see the performance.

307
IDEA-TOPS
艾特奖

获奖者/ The Winners
班堤室内装修设计企业有限公司
（中国台湾）

获奖项目/Winning Project
素颜舞者/ Clean Dancer

获奖项目/Winning Project

素颜舞者
Clean Dancer

设计说明/ Design Illustration

本设计元素属于光环境体验中心之办公室空间，整体共有三层楼面，主要设计概念以杯灯造型做为楼层衔接平台，搭接着弧形结构梯将三个楼层串连。一层为接待大堂及光环境体验展示场与会议空间。二层为主要办公空间及光学研究室。三层为会客室与董事长办公室。

座落于深圳的工业厂房办公室，是一个框架结构的水泥建筑。梁柱结构分明也方正。因此总体设计顺着方正的结构框架平稳的分布与切割，唯独在贯穿楼层的挑空中央部分以圆弧造型梯成为空间的主要视觉焦点。

入门即见的接待柜台正位于挑空造型梯的侧边，柜台台面及立面以人造石无接缝处理，后面主造型墙面搭配木纹大理石，墙面两侧木格栅门可通往服务办公室。天花板小心翼翼的以排列整齐的灯条做为主照明，尽量减少过多传统崁灯的分布，使空间更为简洁，呈现出线条与面之间的比例美感。

本设计的主要构成在于将原本三个个别独立分开的楼层，整体打通贯穿为一个大的挑空天井，运用各楼层错位的弧形阶梯交错其中，在各自不同弧度的连接下，楼梯像舞者般动了起来，透过行进间移动的视线，展现出各种优美的姿态与表情。当踏上踏阶前往各楼层时，人也同时成为了展演舞台的表演者，为严谨生硬的四方框架办公室空间添增了几分的活跃与精彩。

表情丰富的造型楼梯、因为克服楼高及结构支撑问题，所以运用了杯灯造型（杯灯是此公司的产品之一）做为中央支撑平台，同时一层地板也为了考虑地下室楼板在单点受力问题而架上井字型钢梁。整体空间设计没有多余的装饰手法，只运用了结构本身的造型美，搭配简洁通透的玻璃扶手，呈现出最原本的结构面貌，当行走在各楼层的走道间，皆能透过大的天井看见楼梯结构产生的律动、也能穿越楼梯望向不同楼层所造成的景深美。

在灯光照明设计方面，打破传统办公室空间的灯具照明方式。本项目没有一大堆埋入式的崁灯，没有轻钢龙骨的方型灯组，也没有华丽的美术吊灯。运用的是建筑结构形成的面与线所造成的层次，表现原结构在没有经过任何外加造型与装饰下，将光源隐藏在走道两侧的办公隔间框架上与梁柱结构的框架上，光也就在这样的白色渐层中细腻的表现，创造的是空间与光融合的环境，光融入在结构、机能、空间中，整体的空间形成如同一座大型的灯具，而没有了传统单独的灯具。

当设计者透过本质的观察与使用者机能的分析后，运用与保留原有的结构美，以最单纯、最简洁的方式呈现空间最自然的面貌，就像没有浓妆艳抹的美女般，让人喜欢的往往是那素颜的自然美。

在办公室必要的隔间中，由于通道置于室内格局的中间，光线较为昏暗，因此使用通透的玻璃为墙面，除了能使通道空间明亮外，也可让视线相互穿透。但在天地与墙面皆为大面平整硬质材料的空间里，造成了反射的音场与回音是必需克服的问题。部分的运用木质格栅分布在其间，不但能有效造成空间局部的区域分界，更能对全反射的音场空间形成扩散作用，有效的解决回音残响问题。

为了干净的保留原有天花板结构不做任何的多余修饰，玻璃隔间深色的框架及梁柱框架除了支撑玻璃与内藏灯光外，框架上方的隔间墙内成为机电线路与空调管线的分布区，所有传统藏于天花板内东西全都移到了隔间墙两侧，使天花板上及梁上四周没有任何对象的干扰，素颜的表现空间，除了展现最自然的空间表情，也让装修经费达到最精简的状态。

This design element belongs to the office space of a light environment experience center. There are totally three floors. The main design concept is that a cup lamp modeling is used as the floor connection platform to connect the three floors by arc stairs. The first floor contains reception lobby, light environment experience showground and meeting space. The second floor mainly contains office space and optical research room. The third floor mainly contains reception room and president office. The industrial plant office located in Shenzhen is a cement building of frame structure. Its beam and column structure is distinct, upright and foursquare. Hence, the overall design follows the upright and foursquare framework for steady distribution and incision. Only the open-to-below central part running through the floors adopts arc stairs which are the highlight of the space.

There is a reception counter once you enter the door. It is located besides the arc stairs open to below. The top and faade of the counter is seamlessly processed with artificial stone, while the main modeling wall at the back is processed by combining wood grain marble. The wooden grilled doors at the two sides of the wall are accessible to the back service office. Light bars are carefully arranged on the ceiling for main lighting. Traditional recessed lights are minimized to make the space more concise and present an aesthetic perception of proportion between line and surface.

The three isolated floors are connected by an open-to-below patio. Via staggeredinterleaving, the arc stairs are vivid like a dancer and show different elegant postures and expressions along with your movement. When you step on the stairs, you will be like a performer on the stage, which adds some active and splendid elements to the rigid square office space.

The stairs adopt a cup lamp modeling as the central support platform (cup lamp is one of the products of the company) to overcome the problems concerning building height and structural support. Well-shaped steel beams are set up on the first floor to overcome consider single-point stress of the slab of the basement. The overall space is designed concisely and only the modeling beauty of the structure itself is used, matched with concise transparent glass handrails to present the original structure. When you walk in the corridor on each floor, you can feel the rhythm

generated by the stair structure through the big patio and enjoy the depth of field formed by different floors through the stairs as well.

In lighting design, the lighting method of traditional office space is abandoned. This project does not adopt a heap of recessed lights. Neither square lighting sets of light steel keel nor magnificent artistic droplights are adopted. The light sources of the project are hidden in the office partition framework and the beam and column structure to express the sense of depth formed by line and surface without additional modeling and decoration. In this way, an environment blending space and lighting is created, where lighting is integrated into structure, function and space. It seems that the overall space forms a big lamp and there is no traditional isolated lamp.

After texture observation and function analysis, the designers use the original structure to present the most pure, concise and natural space. Just like a beautiful girl without make-up, people usually like her natural beauty.

Since the rays in necessary office compartments are dim, transparent glass is adopted on the wall to make the corridor space bright and the sight interpenetrating.

In a space with ceiling, floor and wall of large flat hard materials, the reflected sound field should be overcome. Therefore, wooden gratings are adoptedin the project to partition local space, diffuse reflected sound field and solve echo.

To retain the neatness of the original ceiling and make no surplus decoration, the dark framework of the glass compartment and the beam and column framework are removed of support glass and build-in lamps. Inner the partition above the framework are distributed with mechanical and electrical lines and air-conditioning pipelines. All the objects traditionally hidden in the ceiling are removed to two sides of the partition. As a result, there is no interference to the ceiling and beam. Simple color expression makes the space mostly environment-friendly and save decoration expenses to the largest extent.

313

IDEA-TOPS
艾特奖

获奖评语

获奖作品在整个建筑物内创造出了一种开放和光明的感觉。空间产生了一种光亮，清澈的和可能性的感觉。

The winning entry creates a sense of openness and light throughout the building. The space generates a feeling of brightness, clarity and sense of possibility.

IDEA-TOPS
艾特奖

NOMINEE FOR BEST DESIGN AWARD OF LIGHTING SPACE
最佳光环境空间设计奖提名奖

DuccioGrassi（意大利）

获奖项目/Winning Project

Max Mara精品北京店
Max Mara boutique Beijing façade

设计说明/ Design Illustration

MaxMara北京精品店的豪华外观并不是店面的外皮。它甚至称不上是一座建筑物。
明亮的双维表面让人无法把握表面之后的任何"形"或体。
它就像是一个光"帘"，从下至上颜色渐淡；就像是矗立于街道与MaxMara世界之间、存在于现实与梦想之间、横间于存在与欲望之间的一道隔膜、一层薄膜。
削弱实体形象，而烘托光线组成的白日梦般的世界：透过明亮的雾云，可能一睹情感天地。
闪亮的不锈钢框架支撑着立面上安装的外层玻璃，并承载着许多纵横交错的不锈钢玻璃板条，将镶嵌在框架上方和下方的LED灯光弥漫开来。
板条的配置十分独特，能够欣赏对面的景色：从里面可以看到外面，反之亦然。
建筑物的结构件均覆盖着反射镜，用于放大效果。

The façade of the MaxMara store in Beijing Avenue Deluxe is not the outer skin of a shop, and even less of a building.

The bright bi-dimensional surface does not let grasp any "shape" or volume behind.

It is a "curtain" of light that fades while ascending, a diaphragm, a membrane between the street and the MaxMara world, between reality and dream, existence and desire.

The physical aspect is diminished in favor of a daydreaming world made of light: a bright foggy cloud, beyond which it's possible to glimpse an emotional world.

Description:

A shiny stainless steel frame supports the façade's outer glasses and bears many slats of satined glass oriented in different ways, that diffuse the light of LED stripes embedded in the frame above and below.

The disposition of the slats allows to see the other side: from the inside it is possible to see the outside and vice-versa.

The structural parts of the building are covered with mirrors, to amplify the effect.

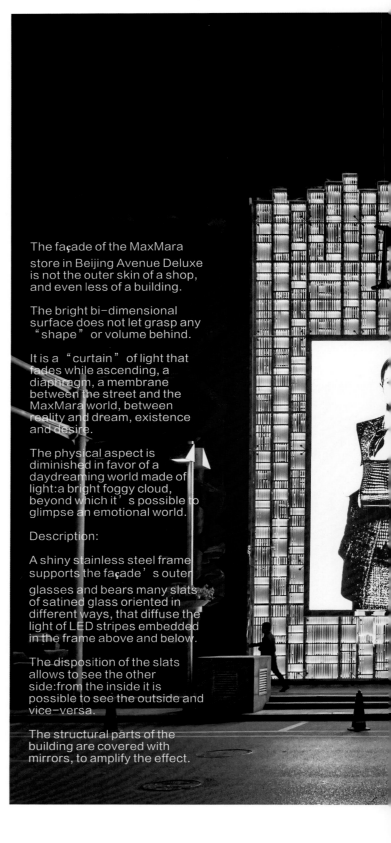

axMara

NOMINEE FOR BEST DESIGN AWARD OF LIGHTING SPACE
最佳光环境空间设计奖提名奖

李辉（中国沈阳）

获奖项目/Winning Project
北京朴贤苑会所设计
Puxianyuan Club

设计说明/ Design Illustration

项目由一层的品酒中心和地下一层的会所两部分组成。

在设计之初，室内设计师就希望能够在有限的空间内营造远离都市的闲逸气氛，同时消除会所本身作为地下空间带给人的压抑感。因此，本案的照明设计师通过运用光影的渗透与叠加，在有限的空间范围内延展设计师赋予的人文气质和悠远意境，达到微妙和谐的文化氛围。

位于地面以上的品酒中心，为了更好地呈现光效，设计师采用幔帐遮挡了自然光的照射，以保证所有的照明都由设定的调光场景完成。室内设计师将最为原始的建筑材料——红砖，作为最主要的立面表达形式加以铺陈。了解了室内设计师的理念后，照明设计师采用单颗1W的LED埋地灯排列在红砖装饰墙面下方，向上的光因为砖墙的粗砺、凹凸质感而产生相互叠加、交错的影，光影融合，质感细腻，沉淀的时光意味得以淋漓尽现。体验店的天花部分则设置了三列不同光束角的轨道射灯，从各个角度对陈列的酒品和金属酒架进行重点表现：深色玻璃的红酒瓶吸收了一部分光线，散发幽隐的光泽；不同釉色的白酒瓷瓶反射了大部分光线，晶莹润泽；酒架的金属本色经过照射，又在天花上反射出具有韵律美的纵横光斑。砖墙的粗糙质感与酒品的光泽质地，一暗一明，一粗一细，形成了戏剧性的对比，相映成趣。恰当温和的光影效果在室内营造了静谧舒适的环境氛围。

通过楼梯步入地下的会所区域，中式的格窗在照明设计师的表现下不再低调含蓄，大面积的背光处理呈现出强烈的欢迎感与仪式感。拾级而下，渗透的光线并不给人以进入地下空间的压抑。楼梯尽头的一组柜体，陈列着各种珍贵的年份酒，被灯光照亮，如艺术品般光泽闪耀，端庄呈现，将主人喜好尽得显露。这组柜体的灯光不但与楼梯的背光在亮度上拉开梯度，同时也为会所内部含蓄隽永的照明气氛铺垫了前奏。

推开楼梯尽头的暗门，是会所的品酒室。室内设计师既保留了这个原本废弃的空间的原始意味，红砖墙面、原木梁架、水泥地面……也赋予了轻奢典雅的格调，丝绸桌布、金属烛台、玻璃器皿……淡淡的灯光从下方照亮红砖墙和原木梁架，粗糙的表面产生出许多的影；原木梁架上的窄角度射灯照亮主桌面，丝绸、烛台及器皿反射出细腻的光泽。在这个空间中，用低亮、暖色温统一粗旷的一面，又以高亮、中色温强调精致的一面，极具戏剧化的光的对比让本来看似简陋的空间反倒散发出经典的韵味，此处为再好不过的品茗红酒并感受其魅力的地方。

穿过悠长的发光走廊，KTV的区域内，窗格背面的云石上连绵的竹林投影图案围拢了整个空间，与楼梯处同样的背光处理手法，在这里通过亮度调节变得朦胧含蓄，发光的桌面兼顾了使用功能，融合在幽幽的竹林之中，为客人营造了一处私密优雅的娱乐环境。

会所内部各个包房内均选用了不同风格造型的吊灯作为桌面的主灯。在场景设定上，照明设计师将其调暗，并不作为桌面菜品及桌花的功能照明。为了保证桌面照度和用餐要求，设计师在天花上补充了窄光束的低压石英灯，单独为桌面补充照度。在包房内的休息区，充满意境的中式窗格外，藤萝薜荔、假山堆叠。照明设计师运用冷暖光色T5荧光灯管阵列和透光膜结合的手段，在"窗"外制造了一方人工天幕，随时间模仿天光的变化，着意于消除客人置身于地下的不安感，同时开阔了包房内的意境。天幕的设计，在施工中也留下了一些遗憾，T5光源与透光膜的距离由于现场的条件限制，过于狭小，使得在某些角度能看到天幕上冷暖相间的光条，成为设计上的遗憾。

与包房连通的铁板烧房，照明设计师把表现的主要方面集中在铁板烧的操作区，巨大的"灯笼"内部分别设置了上下两条灯带，根据不同的用餐需要设置了或通明或渐退的照明效果；桌面下方散逸的光线无形的连贯了周围的座椅，营造多人用餐时的交流感，同时也呼应了天花部分的发光云石和明亮的烹饪区域，避免产生头重脚轻的视觉感受。

作为重要的点景，会所内部的两处水景在照明处理上有意拉开了明暗对比——其一位于包房内，水下设置的灯光照亮水面上安置的金属太湖石，凹凸不平的金属表面流光四溢。为了强调光的存在，照明设计师建议放养了一些锦鲤，有人走过时，鱼群游动打破水面的平静，波光随涟漪投射在天花上，一荡荡无声划动的光，在宁静素雅的包房中游弋；另一处水景位于走廊，连续式洗墙灯由上至下，轻描淡写的灯光在这里更多表现的是影，拼贴的假山在光影下表现出了厚重的质感，与水融为一体，树木的虬枝在视线里留下剪影，以静态的"暗"给人想象的空间，烘托水墨画般的意境美。朴贤苑的照明设计，延续了室内设计师闹市取静的设计初衷，在最终效果上也得到了业主的肯定，正如酒行会所的名字"朴贤"一样，用简单的方式表现室内空间朴素的美，营造舒适恰当的照明氛围，烘托出喧嚣都市内一缕宁静致远。

说明：室内设计方为北京集美组装饰工程有限公司。

This project is composed of the wine-tasting center on the first floor and the club on the first basement.

At the beginning of the design, the interior designers hope to develop a leisure atmosphere far away from the downtown within a limited space. Meanwhile, they hope to eliminate the sense of oppression of the club which is located underground. Therefore, the illumination designers of this case, by applying the permeation and superposition of light and shadow, create a humanistic environment and a remote artistic conception within a limited space and achieve a subtle and harmonious cultural atmosphere.

As for the wine-tasting center on the first floor, the designers adopt curtains to shelter from natural lights for better lighting effect, in order to

assure all illumination is made in the set dimming scenario. They adopt red bricks—the most original building materials—as the main facade expression form. According to the philosophy of the interior designers, the illumination designers adopt 1W LED underground lamps that are arranged beneath the red brick decoration wall. Due to the coarseness and unevenness of the brick wall, the upward lights produce overlaid and interlaced shapes. Integration of the lights and the shadows presents the precipitated times in an exquisite way. The ceiling of the experience shop is set with three rows of track spot lights with different beam angles which are used for expressing wines and wine racks from different angles: the red wine bottles of dark glass absorb a part of rays and emit a hidden gloss; the white wine bottles of different glazing colors reflect most of rays and are glittering and translucent; the metal color of the wine racks reflects rhythmic vertical and horizontal light spots via illumination. The coarse texture of the brick wall and the gloss texture of the wines, dark and bright, thick and thin, form a dramatic and delightful contrast. In this way, a tranquil and cozy indoor atmosphere is developed by the proper effect of lights and shadows.

When stepping into the underground club area, you will find the Chinese-style lattice window will not be low-key and implicit under the expression of the designers and the mighty backlight processing demonstrates a strong sense of welcome and rituality. Besides, you won't feel repressed when stepping into the underground space. A group of cabinets at the end of the stairs display various age liquors with glossy shine and dignified temperament, manifesting the owner's preference to the largest extent. The lights of such cabinets differ in luminance gradient with the stairs, and meanwhile, lay a prelude for the implicit and meaningful lighting atmosphere inner the club.

After entering the blank door at the end of the stairs, you will find the wine-tasting room of the club. The interior designers retain the original characteristics of the discarded space, such as red brick wall surface, log beam frame and cement floor. They also endow the space an elegant style, such as silk tablecloth, metal candlestick and glassware. When faint lights illuminate the red brick wall and the log beam frame, the rough surface will generate a lot of shapes; when the narrow-angle spot lights on the log beam frame illuminate the main desktop, the silk, candlestick and glassware will reflect fine gloss. In his space, low brightness and warm core tone coexist with roughness, while high brightness and neutral color tone highlight delicacy. The extremely dramatic light contrast makes the seemingly humble space give off classical charm as an exclusive place for tasting red wines.

The KTV area comes after the long luminous corridor. On the marbles at the back of the pane, the continuous bamboo forest project pattern surrounds the whole space. With the same backlight processing technique as the stairs, the brightness control here makes the atmosphere hazy and implicit. The luminous desktop considers the use function and is combined in the faint bamboo forest, which creates a private and elegant recreation environment for guests.

Each compartment in the club adopts droplights of different styles as the main lights of the desktop. In scene setup, the illumination designers turn down droplights which do not serve as function lighting for desktop dishes and table flowers. In order to meet desktop illumination and dining requirements, the designers replenish narrow-beam low-voltage quartz lamps on the ceiling. In the resting area inner each compartment, in addition to the Chinese-style pane full of artistic conception, wisteria and climbing fig pile over the artificial hill. The illumination designers adopt cold and warm T5 fluorescent tube array and transmitting film to create an artificial backdrop outside the "window". When i

imitating the change of skylight and eliminating the uneasiness of guests, the artistic conception of the compartment is broadened. The design of the backdrop has some regrets in construction, because light bars with cold and warm tones can be seen from some angles due to the site limitation of the distance between the T5 light source and the transmitting film.

As for the teppanyaki room connected with the compartment, the illumination designers give emphasis on the teppanyaki operation area. Inner the huge "lantern" is designed with two lamps from top to bottom with well-illuminated or fading luminous effects according to different dining atmospheres; the rays dissipated beneath the table make the surrounding chairs coherent and develop a sense of communication, responding to the luminous marbles of the ceiling and the bright cooking area and avoiding top-heavy visual perception.

As important point views, the two waterscapes inner the club feature light and dark contrast in illumination. One waterscape is located in the compartment, wherein lamplights are set to light up the metal Taihu stone and make the rugged metal surface glossy. For emphasizing the existence of lights, the illumination designers suggest keeping fancy carps to develop a dynamic lighting atmosphere; another waterscape is at the corridor, with continuous wall wash lights from top to bottom to highlighting shapes. The stitching artificial hill here manifests a profound texture and integrates with water. The branches of trees leave their shadows in the sight, give people the imagination space and show a beautiful wash paintingconnotation with the static "dark".

The illumination design of Pu Xianfan embraces the original intention of the interior designers—seeking tranquility in downtown. Adopting a simple method to express the simple beauty of the indoor space, Pu Xianfan develops a cozy and proper lighting atmosphere that is tranquil in downtown. Therefore, the ultimate effect is recognized by the owner

Interior designer: Beijing Newsdays Decoration & Construction Co., Ltd.

NOMINEE FOR BEST DESIGN AWARD OF LIGHTING SPACE
最佳光环境空间设计奖提名奖

杭州设谷装饰设计有限公司
（中国杭州）

获奖项目/Winning Project
巴鲁特男装轻奢生活馆
绍兴柯桥万达店
Men's Light Luxury Shop

设计说明/ Design Illustration

巴鲁特男装轻奢生活馆是由来自杭州设谷装饰设计有限公司的主创设计师谢银秋先生和徐梁先生共同创作的。项目位于绍兴柯桥万达广场一层。黑灰的旋律在这个空间里相互交织，硬朗的线条、裸露的材质、夹带着蓝绿色的光源，不仅有着丝丝的调皮与时尚，也叫嚣着巴鲁特的独一无二。整个空间设计，是巴鲁特服装的延伸和续写，设计师采用原始的工业原料，水泥与钢筋的碰撞，硬朗的风格处处彰显着男士的沉稳与内敛。

Brloote Men's Wear Affordable Luxury Life House is jointly designed by chief designers Mr. Xie Yinqiu and Mr. Xu Liang from Hangzhou Shegu Decorative Design Co., Ltd. The project is located on the first floor of Wanda Plaza in Keqiao District, Shaoxing City. The gray black melodies intervening in the space, firm lines, exposed materials and bluish-green light sources with few naughty and fashion elements claim the uniqueness of Brloote. The design of the whole space is the extension and continuation of Brloote apparel. Original industrial raw materials including cement and rebar are adopted by the designers to form a firm style for manifesting the calmness and restraint of men.

326

IDEA-TOPS
艾特奖

NOMINEE FOR BEST DESIGN AWARD OF LIGHTING SPACE
最佳光环境空间设计奖提名奖

上海长乐设计工程有限公司
（中国上海）

获奖项目/Winning Project
贵州非物质文化遗产体验中心
—— 黔元傩
Non Material Cultural Heritage Experience Center

设计说明/ Design Illustration

回眸千年——"贵州非物质文化遗产体验中心"项目简介

非物质文化就像一张薄纸，而社会的发展就像一列高速飞驰的列车，这张薄纸飘在窗外，只要一把抓不住，它就会"唰"地一声飞得无影无踪，再也无处可寻。

贵州作为一个多民族的省份，有着丰富的非物质文化遗产亟待拯救与保护。而非遗保护，既需要合适的土壤，更需要合适的载体。长乐集团选择在一个由古建筑构建的特殊空间中承载起非物质文化遗产的保护、传承与传播，无疑是一个绝妙的选择。

贵州非物质文化体验中心，由三栋从各地迁移而来的名人故居组成，分别是：始建于清康熙年间的清代中兴之臣丁宝桢的长子丁体常的布政使公署；始建于清嘉庆元年，著名外交家和散文家黎庶昌的私宅"英华楼"，门额上"英华钟秀"四个大字乃曾国藩亲笔所书；以及始建于清康熙年间的"都统楼"，房主人是清军的副都统谷丰。

在园区中心水景的两侧，三幢错落的徽派古宅一改传统的粉墙黛瓦，经过写意变形，保留了马头墙的神韵，现代感的线条刻画得却是古韵盎然。

在这三幢古宅中，分别布局了黔元傩、黔学院和黔香阁三个不同的功能空间，集古建筑、非物质文化和非遗菜肴于一体，通过最先进的技术手段和最时尚的方式让来者在视觉、味觉、听觉、嗅觉各个方面体会中国传统文化和历史，从而也赋予这些古老原始的非遗文化以全新的生命。

在黔元傩的中庭，一张12m的长桌可让人明显感受到来自贵州民间长桌宴的气氛，它可同时容纳36人用餐。当24台专业投影机和20块升降式幕布全部打开时，演员和观众瞬间全都变成了剧中人和戏中景，其震撼感觉超过现在的4D球幕电影。在可以多角度旋转的古戏台上，常年上演由中国当代著名戏剧导演陈薪伊主创的薪傩戏《人之太初》。

上个世纪80年代，被誉为"中国戏剧的活化石"的傩堂戏在贵州思南等地被发现。傩戏，是一种内容庞杂的综合性原生态文化事象，它祭中夹戏，祭中有舞，戏在祭中，融合了宗教、祭祀、巫术、文学、舞蹈、戏剧等多种文化元素，中国戏剧也正是从中破茧而出。2003年，傩戏被列入第一批国家级非物质文化遗产名录，而面具是傩戏最引人注目的显著特色之一。

《人之太初》结合现代多媒体技术，将古典与现代巧妙地对立统一，构成了神秘深邃的美感。光、色、影、电、声的从容调度更加深刻地展现出了丰富多彩的贵州民族文化。

戏台大幕引进德国多玛科技，3m×5m的大门可以自由平移，还可以当活动布景，两扇5m高的侧门，可以垂直折转，任意角度变化，使整个舞台充满动感和奇幻。数十台投影仪可以生风、造雨、流火、喷水，让人目不暇接。

黔学院是一个展示空间，展示的是贵州地区国家级和省级非物质文化遗产，苗族银饰和面具通过像素移位、透视重叠的手法，产生裸眼3D的效果，充分展示出少数民族民间手工艺艺术品特有的神韵；采用不锈钢镶嵌工艺制成的"贵州文化江山墙"从北到南、从东到西全面展示十七万八千平方公里的贵州民族风情、民俗文化、自然风光和悠久历史，这种全景式的展示艺术充满了蒙太奇的魅力；展厅中的贵州历史名人走廊，展出了两千多年来为贵州灿烂的历史文化做出杰出贡献的几十位历史文化名人的光辉业绩。此外，还在这个空间展示了中国十大名陶之一贵州平塘牙舟陶和苗绣、丹寨造纸术等非遗实物。

黔香阁，是传承非遗美食的精菜坊。进入厅堂，中式古民居与现代建筑材料的融合使整体散发出文化和典雅的气息，西式家具在中式厅堂的照耀下，显得复古时尚，就连洗手间的设计也独具匠心，门前洗手台上树立的而圆镜可以360°旋转，洗手台的顶花图案是由大理石片组成的艺术中国结，就连标识也是有着几千年历史，当今世界上唯一还在使用的象形文字"水书"。

红色的感应式移门是贵州大方漆器的制作工艺，在明、清时期，大方漆器就被选作"贡品"上京供奉帝王。这里的每扇门背后的图案都是贵州省的省花——杜鹃花，而每扇门的正面都是艺术家手工绘制的各种民族风格的图案。这些具有地方特色的传统工艺美术作品，增添了"黔香阁"古色古香的雅致。

作为中华民族的根与魂，非物质文化遗产是我们民族的DNA，"保护文化遗产，守望精神家园"已成为全社会的共识。贵州非物质文化体验中心对古建筑异地重建，与传统文化嫁接，与现代元素和先进技术对接作出了革命性的尝试。

古建筑在应用过程中，只有与人发生关系，与所在的地域文化发生关系，见证历史，见证传承，才能体现它的价值。每一个古建筑都有其恒久的生命力，藏则毁、居则活。贵州非物质文化体验中心的出现，验证了中国古建筑与中国传统文化、非物质文化结合的可能性。唯有这样，才可能让更多的古建筑重新焕发生命力。

人类文明发展至今，传统文化和民间技艺在物质文明的冲击下日渐式微，无数辉煌的非物质文化遗产被岁月的风沙掩埋。非物质文化遗产不仅可以满足人们认知世界、探究历史、了解特色文化的需求，还将作为一种无可替代的精神财富恩泽后世。

贵州非物质文化体验中心，一个回眸千年的窗口。

Review of a millennial history

—Introduction to "Guizhou Intangible Cultural Heritage Experience Center"

Intangible culture is like a piece of thin paper while social development is like a train running at high speed. The thin paper drifts outside the window. If you fail to catch the paper, it will fly without a trace and be nowhere to be found.
As a multiracial province, Guizhou has rich intangible cultural heritages in urgent need of salvation and protection. The protection of intangible cultural heritages requires suitable soil and carrier. Changle Group selects a special space formed by ancient architecture to protect, inherit and spread the intangible culture heritages, which is undoubtedly a wonderful choice.
Guizhou Intangible Cultural Heritage Experience Centeris composed of treeformer residences of celebrities moved from various places, including Commissioner's Government Office of Ding Tichang (the eldest son of "Ding Baozhen", an resurgent official in Qing Dynasty) built during Kangxi period, "Yinghua Tower" (a private house of Li Shuchang, a famous diplomatist and proser) built in the first Year under the regime of Emperor Jiaqing in Qing Dynasty ("Ying Hua Zhong Xiu" on the head casing was inscribed by Zeng Guofan.) and "General's Tower" built during Kangxi period in Qing Dynasty (the house owner was Gu Feng, a vice-general of the army of Qing Dynasty).
On the two sides of the waterscape in the center of the park, three scattered ancient houses with Anhui style do not adopt the traditional white walls and black tiles but reserve the charm of horse-head wall after transformation. The modern liens depict the ancient charm.
In these three ancient houses, Qian Yuan Nuo, Qian Xue Yuan and Qian Xiang Ge were arranged with three different functional spaces to integrate ancient architecture, intangible culture and intangible cuisines as a whole. Visitors may experience the traditional Chinese culture and history from vision, taste, audition and smell through the most advanced technical means and fashionable mode, and these ancient and original intangible culture are endowed with fresh life.

In the atrium of Qian Yuan Nuo, you may obviously feel the feast atmosphere from the long table lengthening 12 meters in Guizhou. The long table can hold 36 people for dining. When 24 special projectors and 20 elevated curtains are all opened, performers and audiences instantly become the dramatis personae and the scene in the play. Its astonishing feeling outdoes 4D circular screen cinema. On the ancient stage with rotatable angles, Nuo OperaRen Zhi Tai Chu created by Chen Xinyi (a famous dram director in contemporary China) is performed all year round.
In the 1980s, Nuo Tang opera" (hailed as "the living fossil of Chinese drama") was discovered in Sinan, Guizhou Province. Nuo Opera is a comprehensive original culture item with complex contents. Play and dance are available in sacrifice. It integrates multiple cultural elements such as religion, sacrifice, witchcraft, literature, dance and drama. Chinese drama was just born from this. In 2003, Nuo Opera was listed into the national intangible culture heritage list. Mask is one of the most remarkable characteristics in Nuo Opera. Ren Zhi Tai Chu combines modern multimedia technology to skillfully unify the classic and the modern and form a mysterious and deep sense of beauty. The deliberate arrangement of light, color, shadow, electricity and sound profoundly display the colorful ethic culture in Guizhou.
DormaTechnology was imported for stage curtain. 3m × 5m gate can be translated freely and used as an action set. The two side doors with a height of 5m can be bent over with any angle changes to make the whole stage dynamic and bizarre. Tens of projectors can generate wind, rain, fire and water, which provides many things to see.

Qian Xue Yuan is an exhibition space which displays national and provincialintangible culture heritages. Miao silver jewelry and masks generate 3D effect for naked eyes through pixel displacement and perspective overlapping, which fully displays the unique charm of folk handicrafts and artworks of ethnic minorities. "Guizhou Culture Landscape Wall" made with stainless steel inlaying process fully displays Guizhou's national customs, folk culture, natural scenery and long history, ranging from the north to the south and the east to the west and covering 178,000 km. This kind of panoramic exhibition art is full of Montage charm. Guizhou historical figures corridor exhibitsthe brilliant performances of dozens of historic culture celebritiesmaking outstanding contribution to the splendid historical culture of Guizhou for over 2,000 years. In addition, top 10 famous Chinese ceramics are exhibited in the space. Pingtang Yazhou ceramic, Miao embroidery and Danzhai papermaking technology in Guizhou are also exhibited.

Qian Xiang Ge is a fine cuisine workshop inheriting intangible heritage. In the hall, the combination of Chinese ancient dwellings and modern building materials release the cultural and elegant atmosphere. In the Chinese hall, the western furniture looks retro

and fashionable. The design of washroom is of great originality. The round mirror erected on the wash basin in front of the door can rotate at 360°. The poppyhead pattern of wash basin is the artistic Chinese knot made of marble pieces. The logo has a history of several thousand years, adopting the pictograph "Shui script" which is exclusively used in the world.

Red inductive sliding door adopts the production process from Guizhou Dafang lacquer. During Ming and Qing dynasties, Dafang lacquer was selected as "a tribute" to serve the emperors in Beijing. The patterns behind every door is the flower of Guizhou Province— "azalea". The front of every door is the pattern with national style which is manually made with the artists. The local featured traditionalartworks add the antique elegance to "Qian Xiang Ge".

As the root and soul of Chinese nation, intangible cultural heritage is DNA of our nation. "To protect cultural heritage and keep watching the spiritual homeland" has become a society-wide consensus. Guizhou Intangible Cultural Heritage Experience Centermakes a revolutionary attempt on rebuilding ancient architecture in another place, grafting with traditional culture and connecting with modern elements and advanced technology.

In application, ancient architecture only relates to people and regional culture. Its value can be reflected only after history and inheritance are witnessed. Every ancient architecture has permanent vitality. If the ancient architecture is hidden, it will be ruined. If it is lived with people, it will be alive. The birth of Guizhou Intangible Cultural Heritage Experience Centerverifies the possibilityof combining Chinese ancient architecture, traditional Chinese culture andintangible culture. Only in this way might make more ancient architecture release the vitality again. As the human civilization develops till now, traditional culture and folk skill fade away under the impact of material civilization. Numerous intangible culture heritage is being buried by winds and sand. Intangible culture heritage can meet people's requirements of cognizing the world, exploring history and learning about the distinctive culture. It will acts as an irreplaceable spiritualwealth benefiting the later generations. Guizhou Intangible Cultural Heritage Experience Center is a window to review a millennial history.

Today Chapeau is among one of the top multibrand stores of Spain. "Stores have to renovate, not only the product but also the spaces."

ARCHITECTURE OF SPACE
A large black steel marquee is illuminated through openings, it is the frame for the glass facade of the store, presenting separate entrances for men and women as scenes for showing the collections. The shop winfdow is usually characterized by the space where mannequins are, in this case is breaking natural barriers and visual boundaries having the entire shop as background.
Inside the store you pass through a sequence of scenes beginning with the access, dominated by two large LED screens

with video art compositions produced by Javier Santaella for Chapeau. The next scene is defined by long pieces that emerge from the walls where you find the stone voulme counters emerging from the ground connecting spaces between male and female areas. Reaching the changing rooms the areas are visually separate again. At the end the space is joined again in a high background topped by a skylight that floods with natural light the interior of the store.

This progression of spaces organizes the store in areas that highlight the identity of the collections in order to give appropriate value to each brand exposed.

The project is defined by pure geometry which is represented by three basiccategories. On one hand, sets of prisms materialized through various items such as boxes of reflecting mirrors in the center. This prims are used as changing rooms. In the other hand white light volumes on the perimeter and large vertical black mirror at the higher space.

The next category is about the planes that form the ceilings and skylight. A succession of inclined planes along a longitudinal path can accommodate facilities without losing height. These planes lead to the end of the shop, where they transform into angled glass sheets that close the skylight on a horizontal plane. The most basic category corresponds to the lines formed by the clothes hangers and the fine pieces of furniture where clothes rest and hang. This element is also found in the openings that invade the whole space, breaking planes and generating edges. The formal language defined by the basic geometries of solid, plane and line,materializes with glass, mirrors, stones and metals in white, gray and black tones to achieve the ideal atmosphere for exhibiting the highest quality product in an attractive play of reflections, shades and lights.

What is worth mentioning is that the water platform, considering the mass space around, has to be dynamic and remain contrasting to the stair, so I created a 4*4*4 meter cubic frame, twisted horizontal and vertical angles so as to bring forth visual echo with the stair. Set a horizontal glass base inside the platform to project light from one sidelike a slant huge containerfilled with water inside.

The performance stage adopts a leisure design, it seems like a kid folded a piece of paper and blew a breath of air to it, then it gradually grew bigger and bigger, finally landed in the hall and attracted crowds of people to see the performance.

获奖者/ The Winners
北京亚禾工程设计有限公司
（中国北京）

获奖项目/Winning Project
和堂叙/Hall of Peace

334

IDEA-TOPS
艾特奖

获奖项目/Winning Project

和堂叙
Hall of Peace

设计说明/ Design Illustration

以古琴为主题

古琴,距今已有3000年的历史,是八音之中的丝,音域宽广,音色低沉,余音悠远,是中国古老文化的标志性符号,所以对琴舍的设计运用了很强的古典情调,同时保留了现代生活简明的浅色调和全现代生活所需。

建筑外墙的瓦片是根据宋徽宗的《瑞鹤图》,徽宗挚爱绘画、崇尚艺术,象征着中国审美的最高境界。而在设计师的理解,一堆瓦,盖在檐上,是压力,贴着墙根,是担当。因为有这种担当,才引瑞鹤栖于屋上。

"应怜屐齿映苍苔,小扣柴扉久不开。"古琴是幽静的乐器,是大隐于市的气质,和堂叙院子里的地面以青砖铺就,夏季雨水多时其上布满青苔,小院是篱笆扎在一起做为院墙。进入院子,有流觞曲水的小池,池上有木质的茅草小亭,亭中有老木墩、石茶盘、老石柱礅,古雅的精美石雕花盆,有隐士风骨。院中云形的石雕潇洒脱俗,与院墙相携成趣,翠竹掩映,别有生机。和堂叙的大门很隐蔽,是需要寻找的,很多人不是特意拜访或极有兴致是找不到门的,唯推开印有古琴形状的白色"墙壁"才能进入。经此一开,别有洞天……

"抱琴观鹤去,枕石待云归"巨大的青石板石桌,被两个唐代的石墩支撑。桌上天光一线,似宣纸被纸刀划开的灯光带,屏风笔法空灵、刚劲,映着景观中的一石一木,近百年的紫薇象征着好运与深情。墙面有一琴一张,琴体与墙面溶为一体,上有漆裂,仿似历经百年的古琴断纹,又似置素琴于大雪莽莽之中。

琴舍的走廊凌空虚坐一尊水月观音,墙面一排竖挂12把古琴,观音为明代造像,指尖轻挽,似抚琴,地面纯白,打磨如水,遥望神像,宛座虚境水中,不觉琴音拂面,尊严有声。走廊另一面是洗手间,洗手台为樟木制,水淋其上,有幽香。水管金制从天而降,状如发簪,上为镜面,望之绵延不绝,镜边如盘簪,肖清代女子妆容而来。一解易经中为"天一生水"的典故,一为女儿梳妆之用,故陈列精致。

琴舍的里间,上置金制飘带,有绵延之姿,灯光错落其间,墙面有山峦倒影,尽头有一老匾,上书"琴心剑胆",生气巍然。

和堂叙的设计,基于情怀,手法与工艺的应用也都源于初心。琴,涵盖了古代的哲学、天文学、医学、文学、社会学等多方面的文化渊源,琴舍的整体选用最多的也是石、木、金的材质,加之经典的黑白色调,墙面做断纹,复原古琴本身的历史感的同时用玻璃满墙、白色地面,和科学的空间切割运用、视觉延伸使之具有时尚的现代感。

Themed as Chinese zither
Chinese zither, with a history of 3,000 years, is known of the silk of the eight tones and has broad range and lingering sound, being a landmark of Chinese zither culture. Hence, the zither house is designed with an emphasis on classical sentiment and with a detainment of modern life elements.

The outer wall is tiled with the Auspicious Cranes of Emperor Huizong of Song Dynasty. Huizong is keen on painting and art, which shows the aesthetic zenith of China. I understand tiles as a pressure when placed above an eave and a support when adjoined to the wall foot, because only such a support can attract auspicious cranes.

"My clog-print on greenish moss must have incurred the spite; I knock at the wooden gate, but there's no answer after all." Chinese zither is a tranquil instrument with hidden temperament. The courtyard of Hetangxu is paved with blue bricks and enclosed by fences. It is overgrown with greenish moss when rains are plenty in the summer. In the courtyard floats a winding cuvette, on which small thatched pavilions are built with ancient timber piers, stone tea trays and ancient stone column piers set therein. The quaint exquisite stone caving flowerpots suggest the strength of character of hermits. The cloud-shaped stone sculptures are natural and unrestrained. They are shaded by green bamboos to well fit the courtyard. With a hidden gate, it is hard for people to enter Hetangxu, for only the white "wall" printed with the shape of a zither is the entrance. Once opened, a totally different world will present in front of you…
"Watch cranes with a zither and rest head on a stone to wait clouds back". The huge blue slab stone table is supported by two stone piers of Tang Dynasty and has a skylight line like the lamplight gashed by a knife on a rice paper. There is a screen with intangible and bold calligraphy, which responds to all elements in the scenery. The crape myrtle about 100 years emblems good luck and deep feeling. A zither on the wall with lacquer cracks is like ancient broken grains and placed in heavy snow.
A Water-Moon Avalokiteshvara with the Ming dynasty statue is set above the corridor. Twelve zithers are put by the wall in a row. The Avalokiteshvara seems playing zither with fingertips. The floor is pure white and polished like water. Seeing afar, the dignified statue is like in water with a zither sound played out. On the other side of the corridor is a toilet with the elements of Qing Dynasty.
The inner room is set with golden ribbons with a stretching shape. Dotted with lights, there seems a chain of mountains on the wall. In the end is set with an ancient plaque written with "soul of zither and courage of sword".

Hetangxu is designed on the basis of feelings. The application of means and techniques is originated from the beginner's mind. Zither embraces the cultural origins of ancient philosophy, astronomy, medicine, literature, sociology and other subjects. The zither house, as a whole, mainly adopts such materials as stone, wood and, good. While with the black and white panorama and the wall with lacquer cracks to recover the historical sense zithers, glass is used to fully cover the wall and the floor is whitened, so that a modern sense is well presented.

获奖评语

设计以中国经典文化符号实体,构建了舒适雅致的生活空间。

The design creates comfortable and elegant living space through the concrete body of classical Chinese cultural signs.

NOMINEE FOR BEST DESIGN AWARD OF ART DISPLAY
最佳陈设艺术设计奖提名奖

几何空间设计机构（中国北京）

获奖项目/Winning Project

体验中心样板间
We Space Experience Center

设计说明/ Design Illustration

"WE+空间"项目隶属正弘集团，地处郑州高新设计开发区核心地带。这是一个定位为"为创客服务的青年交互平台"项目，几栋小户型满足年轻人的租住及创业需求，楼内设计了一个共用的大客厅，提供洗衣、做饭、娱乐、休闲、学业习、医疗等服务。在全案的设计定位上我们提出了：自然、人文、梦想。

自然——主色调上我们选择了棕色系，像卡其布、沙子、泥土和大地一样的颜色，体现自然、突出自然、不违背自然。

人文——"以人为本"主角？谁是主角？我们要的是生活，而生活并不是一个人的，世间所有的人，才构成了世界，也是所有的人，才构成了生活。人为万物之载体，从人引发所有事之出处，之根本！

梦想——Dream。"对于生活我们充满了梦想，充满了渴望。"它是一切正能量的产物、一切辉煌的源头、一切广义的传播，所以我们用了蓝色和黄色来表达这个既小又伟大的名词。

此案是一个地产项目，我们想要打造一个有充分体验感的空间，设计团队考虑了整个空间中的气氛和功能性，体验很关键，有了良好的体验感才会有想拥有在这里有一处属于自己空间的欲望！所谓体验，是体验四感，从嗅觉开始，视觉、触觉、听觉感知这个空间，来体会设计师想表达的设计意图，从生活上升到艺术，再回归生活。这一点很关键，我们所做的空间不是只能看，如果要上一个比喻，那么她就像一个姑娘，她不是一个花瓶，而是能上得厅堂下得厨房的好妹子。

在材料上我们用了一个全新的体系——TEKNAI，它可以做出我们所想要的各种面层且无接缝，我们想让这个空间自然、朴实，所以大量用了清水的面材，去掉过多的墙面装饰，让四壁退后成为背景。英国设计师凯莉赫本（Kelly Hoppen）曾说："设计并不是要让大家看不懂，而是要通过简单的表现去传递深邃的理念"，所以还是要回到我们的初衷——以人为本。我们不仅仅是在本案中有很多与人有关的元素，而是将这个概念贯穿至终，从空间到家具都是这样做的，把体验感体现殆尽，每一件家具都是精心挑选，每一个饰品都是有意义及生命的。

四套样板间用了四种不用的表现手法，我们不想去泛泛的去给他们定义这样的风格如何，而是想让我们的体验客户进到每一个不同空间后都乐在其中的感受，我们认为这才是设计最重要的关键，所以如果你想感受和体验那就来WE+空间，在其中遇见一个完全不同生活的你！

备注：此项目前期设计定位为"CLOUD艺术公寓"，后期开放时项目定位改为"WE+空间"，故效果图和后期落地有一些变化。

"We + space" project, belonged to Zhenghong Group, is located at the hub of Zhengzhou New & High-tech Design Development Zone. It is positioned as a "youth exchange platform serving for enterprising residents". The small family house type meets the rent and business requirements of young people. A big sharing sitting room is designed in the building to provide such services as laundry, cooking, entertainment, leisure, learning and medical treatment. For this case, we put forward the following positioning: nature, humanity and dream.

Nature—We select bronze tones as the main tones, such as the colors of khaki, sand, soil and earth, which embody and highlight nature, rather than going against it.

Humanity—"people oriented". Who are the major players? We want a life for all people, while all people constitute the world, i.e., all people constitute the life. People are the carriers of all things on earth, so that people are the root for all the things.

Dream—"full of dreams and hopes for our life". It is the outcome of all positive energies, and origin of all glory, and the spread of all generalization; hence, we use blue and yellow colors to express this small and great noun.

This case involves a real estate project. We aim at developing a space with good experience. The design team considers the atmosphere and functionality in the whole space. Experience is crucial, because only good experience can make people have a desire of its own space here! For this space, experience can start from smell to visual sense, touch and hearing. The design intention of designers can be experienced from life to art and then back to life. This is very important, because the space we design does not serve for appreciation only; we also play an emphasis on its functionality.

In terms of materials, we use a brand-new system—TEKNAI which can achieves the various seamless surface layers we want. We hope this space natural and simple, so that we use a lot of water plane materials and remove excessive wall decoration, which makes walls become a background after removal. The England designer Kelly Hoppen once said, "design is not to make people cannot understand, while it is to transmit profound ideas via simple expressions." When we mention "people oriented", we not only apply a lot of elements related to people but also use this concept throughout the design, either from space to furniture, in a view to

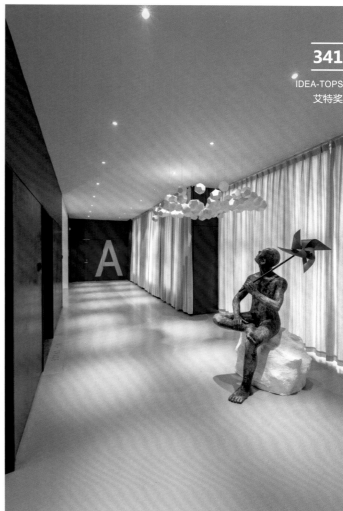

fully embody the sense of experience. Besides, we select all the furniture meticulously, so that each ornament is meaningful and lively.
Four different techniques of expression are used for the four prototype rooms. We aim to let customers experience the sense of different spaces, which we consider is the most important thing for design. If you want to have this kind of experience, please come WE + space to encounter you with a completely different life!

Remark: The earlier design of this project is positioned as "CLOUD art apartment". When the project is opened later, it is positioned as "WE + space", thus there are some changes between the effect picture and the final work.

342

IDEA-TOPS
艾特奖

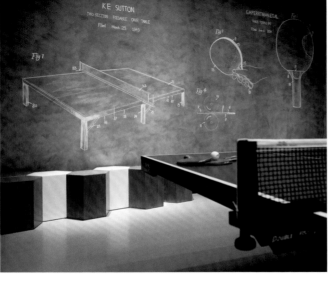

NOMINEE FOR BEST DESIGN AWARD OF ART DISPLAY
最佳陈设艺术设计奖提名奖

深圳市昊泽空间设计有限公司
（中国深圳）

获奖项目/Winning Project

苏州建发地产中浪天成项目售楼处——庄生梦蝶
Dream Of Butterfly

设计说明/ Design Illustration

无论个人或人类的发展都会经历两个过程。第一个过程即人性对动物性的超越，即文明、社会、规则、安全；第二个过程却是对人性的超越，往往体现为宗教或哲学上的形而上，或终极的神性。我个人理解为精神上人性束缚的自由和解放。而在当下社会的剧烈发展和变革中，我们每个人都无一幸免的时刻经受着人生意义的纠解和拷问，茫茫宇宙，何处投路？普遍的精神困局来自于无法对人生现实目的性的超越，即超越功利、欲望、知识等一切的束缚。因为"我执"的无法放下，使这一过程何其艰难。

碰巧读了《庄生梦蝶》的小故事，会意于庄周竟用如此浪漫诗意的智慧追求自由。虽然充满悲剧性的惆怅，但也让人读来神清气爽，希望满怀。作为设计师，我们常常会体悟到语言文字对人的智性的表达是有很多的障碍的，而视觉表达作为一种语境，往往能摆脱这种困境。正好也借用这个小故事的灵感，让每一位来访的体验者都能有各自不同的不可言说的愉悦和放松。当然，我们不敢奢想不敢妄言，它能带给人持久的精神放松，思想的自由……果真能如此，我们倍感欣慰。

每个人都在追求人生的答案。每当我读到下面这段文字，心中便充满透彻和感动。

"南有悬樋，以成清水；近有林，以拾薪材，无不怡然自得。山故名音羽，落叶埋径，茂林深谷，西向晴空，如观西方净土。春观藤花，恰似天上紫云。夏闻郭公，死时引吾往生。秋听秋蝉，道尽世间悲苦。冬眺白雪，积后消逝，如我心罪障。"
——方丈记

The development of both individuals and human beings goes through two processes. The first process is transcendence of humanity to animality, i.e., civilization, society, rules and safety; the second process is transcendence to humanity, which is usually embodied in religious and philosophical metaphysics or ultimate divinity. I understand it as freedom and liberation of mental constraint of humanity. In the rapid development and change of modern society, all people are experiencing the entanglement and interrogation of the significance of human life all the time. To relieve people from the constraints of material gain, desire and knowledge seems very hard. The short tale of "Zhuang Zhou's Butterfly Dream" suggests pursuing freedom with romantic poetic wisdom. Although it is full of tragic melancholy, it makes readers refreshing and hopeful. As designers, we always feel that language and writing have a lot of obstacles on human's wisdom, while visual expression can get rid of such obstacles. The inspiration from the short story enables all visitors to experience their joy and relaxation that are hard to explain. Of course, we hope to give rise to temporary mental digression and thinking freedom…If so, our efforts are worthy.
Each person is seeking the answer of life. We I read the following text, I feel very moved.

"I live by water because I can survive with enough water; I can get warmed by collecting fuel wood from the forest. The valley weeds are so dense that their leaves bury the footpath. Looking west of the blue sky, it seems the heavenly paradise. In spring, there is the overwhelming fragrance of flowering vines. In summer, a gowk twitters and shows me the way. In autumn, I listen to the chirp of autumn cicadas as if they speak of the sadness of the earth. In winter, snow stacks and melts like my inner confusion." —Essays of Idleness

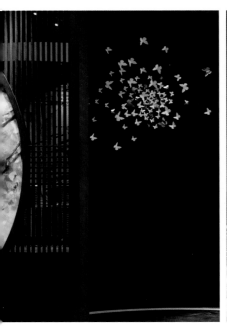

NOMINEE FOR BEST DESIGN AWARD OF ART DISPLAY
最佳陈设艺术设计奖提名奖

DaanRoosegaarde
（荷兰）

获奖项目/winning Project

彩虹车站
Rainbow Station

设计说明/ Design Illustration

2014年12月11日，艺术家DaanRoosegaarde在阿姆斯特丹中央火车站公开了一场大规模的光学艺术作品，荷兰文化和科学技术部长以及阿姆斯特丹的副市长出席。
该艺术作品是为纪念NS和ProRail翻新125年的历史站而发起的。
该艺术作品将建筑师Cuypers建设的历史站与今天的创新连接了起来。
光线和色彩在晚上为参观者提供了一个神奇的时刻。
每年大约有5000万人从阿姆斯特丹中央火车站经过，到达国内和国际的目的地。
全年每天可以在日落一个小时内在车站东边看到"彩虹站"的最美时刻。
DaanRoosegaarde评论道："彩虹站为旅客创造了一个独特的地方，这是一种你不能下载的体验"。
"这种创新和丰富多彩的艺术作品为这座不朽的车站增添了氛围，这也是送给我们旅客的一个伟大的礼物。"ProRail的帕特里克巴克表示和NS的蒂莫·休斯表示。
通过与莱顿大学的天文学家合作，Roosegaarde开发了一种滤镜，可以有效地将光线分解为一系列光谱的颜色。
莱顿天文台的科学家FransSnik表示："由于新的液晶技术，我们正在开发用于研究系外行星（环绕恒星而不是太阳的行星），彩虹站采用了45m宽历史站屋顶的确切形状"。

https://www.studioroosegaarde.net/project/rainbow-station/photo/ #rainbow-station。
该火车站的125周年庆祝活动及其演变过程是Roosegaarde创造其艺术作品的灵感来源。
该项工作也标志着2015年联合国教科文组织（UNESCO）的国际"光之年"。

On the 11th of December 2014, artist DaanRoosegaarde opened a large-scale light artwork at the Amsterdam Central Station, in the presence of the Minister of Education, Culture and Science of the Netherlands and Amsterdam's deputy mayor.
The artwork is launched in honour of the -by NS and ProRail- renovated 125 year old historic station.
The artwork connects the historic station by architect Cuypers with the innovation of today.
The light and colour offers travellers a magical moment in the

evening.
Every year approximately 50 million people travel through Amsterdam Central Station, towards national and international destinations.
Rainbow Station can be seen every day for a whole year for a brief moment within an hour after sunset at the east side of the station.
DaanRoosegaarde "Rainbow Station creates a unique place for travellers, an experience you can not download".
"This innovative and colorful artwork adds atmosphere to the monumental station, it is a great gift to our travellers," said Patrick Buck of ProRail and TimoHuges of NS.
In collaboration with astronomers of the University of Leiden, Roosegaarde developed a lens filter that unravels light efficiently into a spectrum of colours.
Scientist FransSnik of the Leiden Observatory: "Thanks to new liquid crystal technology that we are developing for research on exoplanets (planets orbiting stars other than the sun), Rainbow Station takes the exact shape of the 45-meter wide historic station roof".
Watch a movie of the artwork in action here: https://www.youtube.com/watch?v=LEZCqBIdon0 and
https://www.studioroosegaarde.net/project/rainbow-station/photo/#rainbow-station
The celebration of the 125th anniversary and the metamorphose of the railway station was the inspiration for Roosegaarde to make the artwork.
The work also marks the international 'Year of the Light' by UNESCO in 2015.

IDEA-TOPS
艾特奖

NOMINEE FOR BEST DESIGN AWARD OF ART DISPLAY
最佳陈设艺术设计奖提名奖

许炜杰 Janus
（中国台湾）

获奖项目/Winning Project

香醇记忆
Sweet Memory

设计说明/ Design Illustration

本案位于台北市大安区，一间十余年的十一坪老屋，内部格局方正完整，因此在规划上，先重整空间的秩序，再以简单素材及现代风格，铺陈出最舒适优雅的感受。来到内部，将原有的无隔间格局，透过精心的运作及思考，分隔为半开放的卧眠与公共场域，建构出独立且能互通的交流空间，让视觉落于开放式客厅与餐厅空间时，达到无限延伸的效果。

在客厅家俬与配色选择上，均符合屋主的个性与需求，以带有设计感的红色沙发勾勒公共空间，并以沉稳的灰色漆面作为卧房的床头景色。此外，更保留窗下的原始空间，构置了具有收纳功能的卧榻，一边迎着风、啜饮着咖啡，一边领略场域所给予的真挚温度，使人充分放松并纾解工作压力的空间感。

This case is located in Da'an District, Taipei. It involves a old house of 36.3m2. Its interior layout is square and complete, so that new order of the space is rearranged before applying simple materials and modern elements for achieving the most comfortable and elegant sensation. The original compartment-free layout is divided into a semi-open sleeping area and a public area, so that an independent yet interconnected communication space is created and the sight will be extended infinitely from the open living room and the dining room space.

The furniture in the living room and its color selection conform to individuality and demand of the owner. The red sofa with a sense of design draws the outline of the public space. Gray paint is selected for the bedside of the bedroom. Besides, the original space under the window is retained and a bed with holding capacity is arranged. In face of the wind with a cup of coffee in hand, you will appreciate the sincere temperature brought by the space and enjoy the sense of space of full relaxation and pressure relief.

NOMINEE FOR BEST DESIGN AWARD OF ART DISPLAY
最佳陈设艺术设计奖提名奖

胡勤斌（中国东莞）

获奖项目/Winning Project
中山富元利和豪庭D2样板房
Lihe TFG. Garden D2 Model House Project

设计说明/Design Illustration

东方雅致生活。

当我们置身繁华都市，在喧闹处，有一处静雅天地，大隐居于此，宠辱不惊，闲看庭前花开花落；去留无意，漫随天外云卷云舒。持壶沏茶，品读人生，神交自然，静聆天籁。"所以游目骋怀，足以极视听之娱，信可乐也"。

Exquisite Oriental Life
Settle in the bustling city, whilst the heart longs for solitude, in the process you may undergo ups and downs in life, witness blossom and fade of flowers; where come and leave without a trace, but to let you mind overflow with clouds scud across space. Delight yourself in the art of cooking tea or the perception of life, or nestle in the sight and sound of nature. "This is the great scene that sets your soul free, and awakens your senses to experience true happiness."

357
IDEA-TOPS
艾特奖

360 Best Design Award Of Residential Architecture
最佳住宅建筑设计奖

艾特奖
最佳住宅建筑设计奖
BEST DESIGN AWARD OF RESIDENTIAL ARCHITECTURE

IDEA-TOPS
INTERNATIONAL SPACE DESIGN AWARD

获奖者/ The Winners
Ramon Esteve
（西班牙）

获奖项目/Winning Project
Sardinera别墅/ Casa Sardinera

获奖项目/Winning Project

Sardinera别墅
Casa Sardinera

设计说明/ Design Illustration

Casa Sardinera住宅坐落于西班牙海滨城市阿里坎特，位于一个俯瞰着地中海的陡坡顶端。该住宅由Ramon Esteve设计，由多个白色悬挑混凝土体量构成，面前展现出一望无垠的美丽海景。

木质模板的印迹清晰地留在了混凝土表面上，形成了独特的纹理，从而与房屋内部的白色木质墙面相互呼应。一层包含住宅的主要生活区，在落地玻璃窗前能欣赏到广阔的海景。从这里延伸出去是一个宽阔的露台，最前端有一个棱角分明的大游泳池。

在房子内部的地下一层，有一个室内游泳池，与室外的水池通过一条长条的窗户互相映照，这扇窗户使光线透射到了室内。在上层，设有多间卧室，每间卧室都有小小的玻璃阳台，可欣赏海天一色美景。

Casa Sardinera is located in Alicante, a coastal city in Spain and at the top of a steep hill overlooking Mediterranean. Being designed by Ramon Esteve, it is composed of multiple white overhung concrete masses and shows beautiful boundless seascape. The shape of wooden formworks is clearly imprinted on the concrete surface to form unique texture and echo with white wooden walls inside the house. The first storey is a living quarter. You may enjoy the broad seascape in the front of French windows. It is a capacious terrace by extending here. There is a large swimming pool with clear corner angles at the most significant end.

There is an indoor swimming pool in the ground floor. Through a long window, it can be seen together with the outdoor swimming pool. This window allows the light to come into the room. Each mass holds a bedroom on the upper storey. A small glass balcony is available in the bedroom. You may enjoy the scenery from sea and sky.

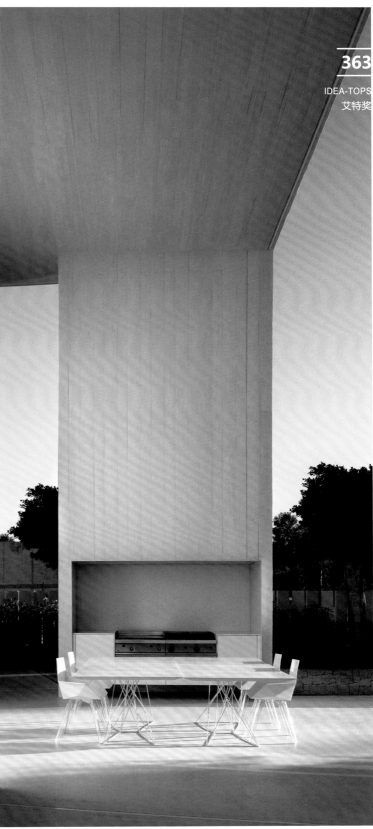

363
IDEA-TOPS
艾特奖

364

IDEA-TOPS
艾特奖

获奖评语

清晰的海景、粗犷的线条和令人心动的地平线都反映在这建筑物清晰而开阔的视野中。

The clear, strong line of the seascape & beckoning horizon are reflected in the building's clarity & openness of vision.

NOMINEE FOR BEST DESIGN AWARD OF RESIDENTIAL ARCHITECTURE
最佳住宅建筑设计奖提名奖

Chang Yong Ter（新加坡）

获奖项目/Winning Project

花果山
Hua Guo Shan

设计说明/ Design Illustration

有对新加坡家庭的祖父母拥有以下三个梦想：他们想拥有一套大房子，以供他们三个已婚女儿居住；他们还想拥有一个果园，果园里种植一些果树、蔬菜和草药，待到收获季节时，便可采摘；此外，他们还想拥有一个良好的生态环境，以供孙子女成长。
当客人达到的时候，便可看到美丽的私家车道。车道两边有果树和灌木环绕。经过瓷砖铺设后的车道显得十分耀眼夺目。
房子重点朝向一侧。房子正面便是果园，阶梯在前面蔓延开来，展现一种迎接来宾的姿态。
在房子平面图中，房间、风景与空间可以相互交替轮换。这种设计可有效地扩大每个房间的知觉空间。每个房间最少有两侧朝向美丽的庭院或空间，因此，房间通风良好，具有良好的透光性，可清楚地看到家庭成员之间的状况，为小孩和成年人提供一种安全感。
房间被安排在不同的水平位置，并用庭院和花盆将其间隔开来。这种设计可以为家庭男女老少提供必要的隐私、互动及和谐的居住环境。
采用钢筋混凝土作为房子的主体结构和建筑形式，为花盆和隔热提供一个完整的不透水结构。
房子规划功能多变，空间可以重复利用，满足灵活多变的需求。
房子屋顶采用阶梯状台式结构，极大地优化了邻里之间的景色布局。屋顶板还起到隔热层的作用，为家庭提供一个更加清凉舒服的环境。
房子重新阐释了热带地区住房的特色。它展示了目前热带地区多代同堂，共居果园的多样需求和愿望。

The grand parents of a Singaporean family wanted a house for their three married daughters; an orchard where fruit trees, vegetables and herbs could be grown and harvested; and an ecological-friendly environment where their grandchildren could grow up in.
Upon arrival, the visitor is greeted by a landscaped driveway lined with fruit trees and shrubs on both sides. This was previously a tiled driveway that was glaring and hot.
The house is tugged to a side. The front façade is a composition of edible orchard, terraced to break down the scale and offers a welcoming gesture.
The house plan is a diagram of alternating rooms and landscapes/voids. This configuration effectively enlarges the perceptual space of each room, as each room opens to either landscaped courtyards or voids on two sides the least.
As a result, the rooms are always well-ventilated and day-lit, with visual connectivity amongst family members, offering a sense of security for children and adults.
By locating the rooms at different levels and using courtyards and planters as separators, this configuration sets a balance in terms of privacies, foster interaction and harmonious living amongst the generations.
Reinforced concrete was opted as the main structural and architectural expression of this house, to provide an integrated and water-tight structure for the planters, and for heat insulation.
This house is also planned for adaptability and reuse, offering flexibilities to changing needs.
The house culminates with a terraced timber-decked roof, optimising the best views of the surrounding neighbourhood. This roof deck also serves as insulators, and contributes to a cooler environment for the neighbourhood.
This house is a re-interpretation of the tropical house. It demonstrates the potential of co-housing an orchard and a multi-generation family serving differing needs and aspirations, in a contemporary tropical setting.

371
IDEA-TOPS
艾特奖

NOMINEE FOR BEST DESIGN AWARD OF RESIDENTIAL ARCHITECTURE
最佳住宅建筑设计奖提名奖

Puran Kumar Design Studio
（印度）

获奖项目/Winning Project

阿利巴格-芒果屋
Alibaug-The Mango House-Home

设计说明/ Design Illustration

芒果屋是追求与自然环境形成联系的物质表现。此处设计的精髓是构思简单并通过结构的形式、材料和装饰进行表达。建筑的有机性质成功将外部与内部相连，从而设法通过其自然流动的规划传递一种泥土般的感觉。房屋融合了"非地区性"的各种元素和建筑材料，使得设计更加简约。

由于芒果树是这一小块土地上的主体，芒果树显然可为房屋赋予一定的含义，以确保发挥作为有机体或所采用绿色文化的基本价值。这片土地上70~80岁的本地居民成为了房屋设计和理念的决定和指导因素。

位于北部、南部和东部方向的芒果树划定了房屋边界。

其目的是为了从房屋中的任何位置处均可观看周围的景观——沿着南北向和东西向轴线。这在所有四个方向开启了一个通道，借此外部青翠的风景一览无余。也作出了一些确定设计——一条朝北的入口，这是因为那里设有车道空间，而厨房则位于东部，以便看到清晨的朝阳。

有必要在开放和覆盖空间之间取得平衡。至于树木对建筑造成的限制，唯一的解决办法是抬高地层，但仍要忠于乡村感。房屋反映出一种自由流动感以及与周围环境的不间断连接，但不会失去设计的比例。

创建体积是结构的一个重要方面，而其具有的斜面屋顶使得结构的最高点大约为35英尺。这是悬浮楼梯的最重要特点，它直接通往上层楼房。此处的天窗和餐饮区的另一扇天窗更加映衬出空间的浩瀚感。

The Mango House is the physical manifestation of a quest to connect with the natural environment. The essence of design here is simplicity in thought and expression through the form, material and décor of the structure. The organic nature of the construction successfully connects the outside with the inside and thus manages to convey an earthy feel through its free-flowing plan. The house is a blend of various elements & building materials that are 'azonic', lending simplicity to the design.

Since mango trees dominated the plot, the house clearly gets its definition from them to ensure that the basic value of being organic or adopting green culture was exercised. These 70-80 year old inhabitants of the plot became the deciding and guiding factors for the design and concept for the house.

The mango trees in the north, south and east directions demarcated the boundary of the house.

The aim was to be able to view the surrounding landscape from any point in the house - along both north-south and east-west axes. This led to an entrance on all four sides for an uninterrupted view of the verdant outside.

There were some certainties that were a given - entrance to the north as there was space for a driveway and kitchen on the east to catch the early morning sun.

There was a need to get a balance between the open and covered spaces. With the restrictions imposed by the trees on the construction, the only solution was to go a level up but stay true to the village feel. The house reflects a free flowing and uninterrupted connect with its surroundings without losing the proportion in design. Creating volume was an important aspect of the structure and with the sloping roof one gets about 35 feet at the highest point. This is most emphatic at the suspended staircase as it sweeps up to the upper floor. A skylight here and another over the dining area underscore the feeling of vastness.

IDEA-TOPS
艾特奖

NOMINEE FOR BEST DESIGN AWARD OF RESIDENTIAL ARCHITECTURE
最佳住宅建筑设计奖提名奖

Masahiko Sato
（日本）

获奖项目/Winning Project

M4[重叠的房屋]别墅
M4-house [overlapping house]

设计说明/ Design Illustration

保护隐私和安全；为家庭创造幸福和安全的空间；充分利用空间。
通过在黑木框上粘贴红杉木，其外观体现了四种不同的节奏层次感。凭借和谐的建筑和周围环境，使人感到温暖和友好。设计师在第一梯级中设置停车场以创造一种带有节奏感的外观。第二梯级中包括通往房子的通道，设计师将其设置在离前方道路稍微偏远的地方。所以，通过这种布置，家人可以通过未指明数目的行人通道保护隐私。在第三梯级和第四梯级分别布置了生活用餐区、配备供水空间的区域，并为每人分配一个房间。
为了充分利用空间，设计师将自行车和汽车的停车空间整合在一起，并且无需使用传统的围墙和车棚。此设计为一个大庭院，可以在其中享受阳光和微风。庭院中种植了树木，将阳台露台与庭院连为一体，这样还能保护隐私。另外，家人也可以在起居室和餐厅感受到季节的变化。
红杉树柱廊为起居室和餐厅增加了一种节奏感。白天，阳光从门廊透过叶片，使客厅和餐厅变得极具友好与和谐感。晚上，借助照射到扶手上的星辰反射光线，让空间变的美丽亲切。
因为天花板外壁贴附有红杉树，使得起居室与餐厅以及整个房子变得更大。通过这样的设计，使得建筑、庭院、家庭和周围的环境和谐统一。

我开始享受这片土地，这里的M4-房屋富含自然元素，而"重叠的房屋"堆积形成的分层则可保护隐私。

Protecting privacy and security; creating a happy and safe space for family; making full use of site. Appearance has four differences in level sense of rhythm by pasting red cedar to black frames. And it makes people feel warm and friendly by harmonizing architecture and surrounding environment.I place a parking lot to the first step of a certain sense of rhythm appearance. The second step included the entrance to a house and located it in the slightly remote place from front road. So, family can secure privacy from the unspecified number of pedestrian traffic by doing it this way. Living dining space and place equipped with a water supply space, a personal each one room are placed to the third step and the fourth step.
To make full use of site, I make parking space for bicycles and cars as a whole without using conventional fences and carport. I design a large courtyard to bring sunshine and wind into interior. I planted trees in the courtyard to make balcony terrace a whole with courtyard and protect the privacy. Also the family can feel the changes of season in living&dining room.
Colonnades of red cedar give a sense of rhythm to living&dining room. In day time, the sunlight going through leaves from atrium makes living&dining room very friendly and harmony. In night, I used indirect light in handrail of stars to make the space beautiful and kindly.
It makes living&dining room and the whole house bigger because the outer wall of ceiling was pasted red cedar. By this design, it also makes architecture, courtyard, family and surrounding environment in harmony with each other.
I become able to enjoy this land where this M4-house is rich in naturally while the layer where "the overlapping house" was piled up protects privacy.

NOMINEE FOR BEST DESIGN AWARD OF RESIDENTIAL ARCHITECTURE
最佳住宅建筑设计奖提名奖

Masahiko Sato
（日本）

获奖项目/Winning Project

N8别墅
N8-house [House of III-BOX]

设计说明/ Design Illustration

这一设计的理念是确保隐私和安全，同时创建一个令家人幸福和轻松的生活空间。

在夜晚，房屋的影子会随着月亮和星星的光线而发生改变，就像人们脸上的表情。

木制的前门散发出一种温馨的氛围，通过此门，人们走进客厅并通过一个巨大的玻璃看到整个院子，给予人们更多大自然的感觉。在隐私花园附近，也就是在客厅的对面，设有厨房和餐厅。

在客厅里，一层楼的天花板上有一个天窗，这让自然光线进入房子。

在楼梯旁的存放处设有一个书房。它包括一个长度为1300mm的书桌。

该楼梯直通到1楼。在1楼里，设有所有家庭成员所用的卧室。

从外面来看，该房子的形状不像通常意义上的2层房屋，它看起来很特别。这一设计充分利用了房屋面积，使得各房间的安排更加紧密，并且能够给予家庭成员之间一个更好的交流机会。

每个房间都分别独立于另一个房间，并且与大自然完美融合在一起。

The concept of this designing is to ensure the privacy and safety, and also create a living space which makes family members happy and easy.
At night, the shadow of houses will change with the light of the moon and stars, just like expressions on people's face.
The wooden front door gives out a warm atmosphere, going through it, one would step into the living room and see the courtyard through a huge class, which gives people more senses of nature in daily life.
Around the privacy garden, which places against living room, there are the kitchen and dining room.
In living room, there is a skylight on the ceiling of the 1st floor, which brings the nature light into house.
Inside kitchen, there is a study placed on the storage besides the staircase. It includes a desk of 1300mm length.
The staircase goes to the 1st floor. On the 1st floor, there are bedrooms for all family members.
Looking from outside, the shape of the house is not like usual 2-floors house. So, it looks quite special. This design make the best use of place, which tighten the arrangement of rooms, it enable a better chance of communication among family members.
Each room is independent of another, and perfectly blending with the nature.

382
IDEA-TOPS
艾特奖

383
IDEA-TOPS
艾特奖

384
Best Design Award
Of Public Architecture
最佳公共建筑设计奖

艾特奖
最佳公共建筑设计奖
BEST DESIGN AWARD
OF PUBLIC ARCHITECTURE

IDEA TOPS

INTERNATIONAL SPACE DESIGN AWARD

385
IDEA-TOPS
艾特奖

获奖者/ The Winners
UNDURRAGA DEVES ARQUITECTOS
（智利）

获奖项目/Winning Project
米兰世博会智利馆
CHILE PAVILION EXPO 2015 MILAN

获奖项目/Winning Project
米兰世博会智利馆
CHILE PAVILION EXPO 2015 MILAN

设计说明/ Design Illustration

本案设计灵感来自于古罗马城市，Herzog and the Meuron通过跟踪战略中心点，设计了2015世博会的总平面图。基于此，他们设立了两条主要道路：Cardo（南北向）和Decumanus（东西向）。沿Decumanus，细长的矩形场地，组织排列得像鱼骨一般。这种"城市布局"启发了设计师设计一个与公共场所密切相关的展馆。在其中，道路就如同传统市场一般在建筑物下方延伸，并笼罩在屋顶的影子之下，但同时与周围环境直接相连。

从一开始，设计师就认为展馆应由木头制成。在智利，木结构是一种美丽、富有内涵的传统，根源于欧洲在美洲的殖民时期。木材也是我国最重要的自然资源之一；它是一种可再生材料，而且智利是世界上退耕还林率最高的国家之一。

为了延长建筑物的生命周期，其结构设计就像一个Meccano（钢件结构玩具），且结构系统简单合理，方便快速拆装各元件，从而能够在2015年世博会结束后将整个展馆拆解、运输并在智利重建结构。同时，其空间设计为中性；不仅在展会期间提供了灵活性，而且能够适应将来的不同用途。

形式上看，该展馆就像一个简单的盒子或容器，其网状结构就确定了要表达的一切内容。结构和建筑是一体的。从远处看，展馆呈一个整体，且规模巨大；靠近看时，结构的复杂性及其组件的尺寸给该建筑物赋予了一种工艺品质和人性。

木箱座落于6根金属支柱上，这种"桥梁"般的形象，完全解放、空出一楼的空间，创造出一种视觉通明度，并允许参观者自由漫步其中。同时，这样的设计方案在城市空间与私密空间之间建立了一种密切关系，缩小了公共与私人空间之间的距离，并融化了此二者之间的界限。

该展馆充分体现了封闭与开放空间的双重性。第一，木质结构下方的开放空间，用作接待公众的门廊。从门廊开始，参观者即开始了一场线性体验。当他们踏上有趣的黑暗活动梯时，便会听到混合着诗歌的岩石开裂声，将他们完全地从外界抽离。在展馆内部，智利展示了其数百万年构造运动塑造的、令人惊叹的地理环境，参观者可经历一场感性的体验。展品结合了智利人民的感情和努力，形成了该国最宝贵的财富果实。

参观者完成了内部贴心亲密的体验后，便可经由嵌入木结构的开放式活动梯下楼。随着活动梯的下移，一张由天然低矮假水青冈木制成的50m长桌，开始逐渐显现在展馆下方。在这里，智利将邀请参观者围绕着这张桌子坐下，与他们分享智利人民的爱和果实。

展馆也可算作是一个小礼堂和一个具有独立通道的多用途房间，在不需要中断常规展览的条件下，可举办临时活动。

Inspired in the traditional foundation of the Ancient Roman cities Herzog and the Meuron designed their Masterplan for Expo2015 by tracing a strategic center point, from which they established two main roads, a Cardo (North–South) and a Decumanus (East–West). Along the Decumanus, long and narrow rectangular lots are organized like a fish bone. This "urban layout" inspired us to design a pavilion closely related to the public spaces where the street extends underneath the building like in traditional markets, protected by the shadow of its roof but open to its immediate surroundings.

From the very beginning, we tough that our pavilion should be wooden made. There is a beautiful and rich tradition of wood construction in our country, which roots are found in the European colonization in America. Wood is also one of our most important natural resources; it is a renewable material, being Chile one of the countries with highest reforesting rates in the planet.

In order to extend the building's life cycle, its structure is designed like a Meccano, where the simplicity and rationality of its construction system allows quickly assembling and disassembling of the pieces making possible to dismantle, transport and rebuild the structure back in Chile once Expo2015 is over. At the same time, its spaces are designed to be neutral; this not only gives flexibility during the exhibition but also allows adapting it to different uses in the future.

Formally, the pavilion is a simple box or container whose expression is defined by its reticulated structure. Structure and architecture is one single thing. From distance it appears as a totality, acquiring a monumental scale; as you get closer the complexity of the structure and the size of its components gives the building a craftsmanship quality and a human scale. The wooden box is seating on 6 metal pillars. This "bridge like" condition liberates the ground floor creating a visual transparency and allowing free stroll of the visitors. At the same time, this strategy establishes a close relationship between urban space and intimate space, narrowing and fusing the line drawn between the public and the private.

The duality between enclosed and open space is reflected in the exhibit. First, the open space under the wooden volume serves as an atrium to receive the public. From there the visitors start a linear experience. They go up through an intriguingly dark moving ramp where cracking rocks noises mixed with poem readings abstract them from the exterior. Inside, visitors go through an emotional experience, where

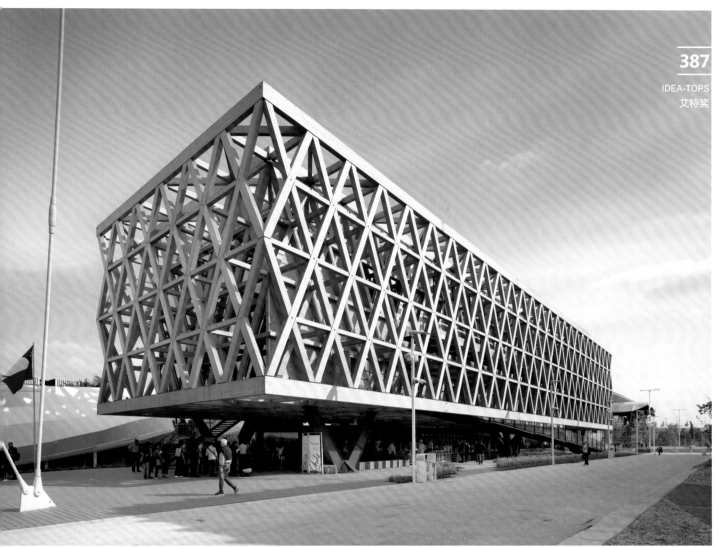

Chile shows its amazing geography, shaped by millions of years of tectonic movements. That combined with the love and effort of the people who live in it, results in one of the country's most valuable treasure, its fruits.

Once the visitors finish the interior and more intimate experience they go down through an open ramp imbedded into the wooden structure. As they descend, a 50 meters long table made of natural Lenga wood, starts revealing under the pavilion. There, is where Chile shares its love and its fruits with the visitors by inviting them all to seat at the same table.

The pavilion also counts with a small auditorium and a multipurpose room with an independent access, allowing hosting temporary events without interrupting the regular exhibit.

获奖评语

有逻辑性、一致性，简单而美丽。2015年世博会智利馆是率性施工和良好坚固建筑的范例。

Logical, consistent, simple but beautiful the Chile Pavillon at 2015 Expo is an example of straightforward construction and good solid architecture

NOMINEE FOR BEST DESIGN AWARD OF PUBLIC ARCHITECTURE
最佳公共建筑设计奖提名奖

清华大学美术学院
Academy of Arts & Design, Tsinghua University

清华大学美术学院
（中国北京）

获奖项目/Winning Project

米兰世博会中国馆
China Pavilion Milan Expo 2015

设计说明/ Design Illustration

摒弃将文化展馆作为广场对象的典型概念，中国馆传达出一种空间领域感。设想为漂浮在"希望的田野"上的一朵祥云，展馆中计划了一系列公共项目，均在悬浮的屋顶下进行。其独特的设计，为项目营造了一个标志性的形象，在世博园区成为一个独特的存在。
中国馆的主题为"希望的田野"。项目主题体现在起伏的屋顶结构形式，南面自然天际线与建筑物北面城市天际线交融的屋顶，表达了城市与自然和谐共存便可实现"希望"的概念。设想为一个木制结构，参考中国传统建筑中的"高架梁"系统，该展馆屋顶还采用了现代科技，创造了适合于建筑公共性的大跨度。参考传统的陶艺建筑屋顶，屋顶覆盖着木瓦面板。但又通过大型竹编板进行了重新诠释，不仅充实了屋顶轮廓，而且遮蔽了下方公共空间。设计为分层遮蔽层，这些面板为展馆屋顶增添了质感和深度，并在下方创造出令人回味的光线和透明效果。

From the architect. Rejecting the typical notion of a cultural pavilion as an object in a plaza, the China Pavilion is instead conceived as a field of spaces. Envisioned as a cloud hovering over a "land of hope", the Pavilion is experienced as a series of public programs located beneath a floating roof, the unique design of which creates an iconic image for the project and a unique presence within the Expo grounds.
The theme for the China Pavilion is "The Land of Hope". The project embodies this through its undulating roof form, derived by merging the profile of a city skyline on the building's north side with the profile of a landscape on the south side, expressing the idea that "hope" can be realized when city and nature exist in harmony.
Conceived as a timber structure that references

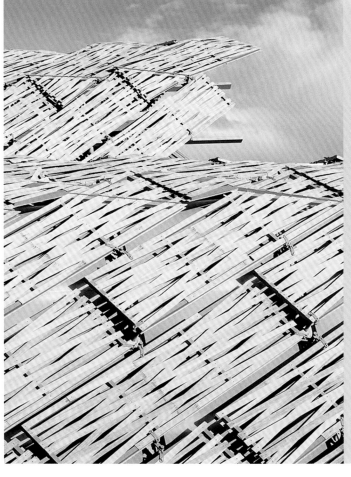

the "raised-beam" system found in traditional Chinese architecture, the Pavilion roof also uses modern technology to create long spans appropriate to the building's public nature. The roof is covered in shingled panels that reference traditional pottery roof construction, but are reinterpreted as large bamboo leaves that enhance the roof profile while shading the public spaces below. Designed as layered screens, these panels add texture and depth to the Pavilion's roof and create evocative light and transparency effects below

NOMINEE FOR
BEST DESIGN AWARD
OF PUBLIC ARCHITECTURE
最佳公共建筑设计奖提名奖

EugeniuProdan（摩尔多瓦）

获奖项目/Winning Project

2015年米兰世博会摩尔多瓦馆
Moldova Pavilion for Expo 2015 Milan

设计说明/ Design Illustration

2015年世博会的主题是"滋养地球、生命能源"，暗示着地球是我们的家，通过均衡营养和安全能源，生活在这个星球上的人们拥有相同的权利，并且我们每个人都拥有平等的机会来获得优质食品、干净水源以及健康的生活方式。
探索纯能量，享誉国际的艺术家Pavel Braila提出将摩尔多瓦共和国的整个展馆改造成一个令人印象深刻的环保装置。基于Gorgona中标方案的展馆建筑风格，Braila建造了一个雄伟的特定场域太阳能装置，使得摩尔多瓦馆可内外同时工作，并让参观者在进入展馆之前就可以感受到这一点。
"太阳能之花"——展馆的关键要素，是一个由无数小平面组成的三重镜面立方体，安装在玻璃房上面一根倾斜金属杆的顶端。在玻璃房各个角落安装的镜面系统会整日跟随太阳的运动，并将阳光反射到旋转的镜面立方体，营造出无数光斑环绕整个空间的效果。由于展馆的透明结构，参观者可在展馆内壁看到这种神奇的效果，离展馆较远的参观者也能在通道、阶梯和草坪上欣赏到这一效果，从而使每一个路过的人都能感受展馆的美丽。镜面装置也配备了太阳能电池，可以为夜间储存能量。所以展馆可全天候运转。
建筑的透明性和令人兴奋不已的太阳能效应（太阳是世界上最大的免费能源来源）吸引观众前来，并且显然是在探讨全球社会问题、环境遗产、开放性以及平等。除了壮观的景象之外，遍布空间的反射光示范了一种转变：是时候重新考虑传统能源的使用、关注环保新能源、保护自然资源以及反对电力资源垄断了。展馆的设计亦借用了普罗米修斯（Prometheus）的神话故事。普罗米修斯从神那里偷来了火种，并将火种交给了人类，为人们带来了启迪和智慧。

长长的走廊一直穿过玻璃房，覆盖着真正的绿色植物，反映了建筑师对环境、自然、人类的关注，并带领参观者进入一个摩尔多瓦传统农业的生态农场区域，这里有各种未被西方世界所熟知的生态产品和著名的葡萄酒。在这里，鼓励人们有秩序地慢慢品味这个展馆所要呈现给大家的东西。进入展馆后，参观者首先看到的是一个特殊的天体现象——在黑色塑料上由霓虹灯做的12星座，用Pavel Braila的话说："这些星座世人尚不知道，只有在摩尔多瓦才能看见。"新娘、农夫、葡萄、奶酪、白鹳、大白鲨等星座……早期的航海家们利用星星为他们引路，发现了新大陆。在这里，星星将引导大众去发现这个有些参观者从未听说过的国家。
艺术家说："当夜幕降临，夜空中群星网罗密布，看着星星，让我想要为你们讲述我的故事，以我自己的讲述方式，讲述一个过于远离现实、而又更接近真相的故事……"人的能量——生命的能量进入主展厅，观众会发现大型投影放映的电影"JOC"，没有对白也没有旁白。为2015年米兰世博会而特别制作，该电影讲述了传说中的民间舞蹈乐队JOC，摩尔多瓦历史最悠久、最传奇的乐队。该乐队成立于1945年，在其悠久的历史长河中，JOC在全球举行了超过7000场演出。该电影歌颂了摩尔多瓦正宗的舞蹈、善良美丽的人民、伟大的音乐和民族服装。Pavel Braila在这部片中制定的影像叙事，以大量的细节描写和令人着迷的乐章，令观众仿佛变成摩尔多瓦人，通过舞蹈的感官和情感的语言，与摩尔多瓦亲密接触。舞蹈和纯粹的喜悦是摩尔多瓦人的核心要素，也是生命中伟大精神和能量的体现。

滋养地球
走出展馆，参观者将通过一个食品角，温馨的氛围让每个人都愉快地品尝纯生物产品制造而成的小吃和饮料，提供的食品都不是转基因食品，而是在自然友好条件下生长的作物。参观者不仅有机会品尝摩尔多瓦的传统美食，还能通过一种新颖有趣的方式了解酒吧产品的培育过程以及摩尔多瓦的农业，并在此过程中显示：虽然有机可持续食品和烹饪在未来将是欧洲和整个世界的选择，但是在摩尔多瓦，这已经是当前和日常的生活；并且摩尔多瓦已经准备好与世界分享我们的一切。

具体侧面效果
每一届世博会均是一台巨型机器，代表了世界上的各个国家，并且每个国家都愿意在世博会投入大量的资金。从这个意义上说，米兰世博会也不例外。但是可能是因为本届世博会的主题为"滋养地球，生命能源"，所以让所有的参与者在开支花销、环境污染和材料浪费方面花费了不少心思，并进行另类思考。展馆只是现代的结构，通常在展会后几乎所有的展馆均将摧毁或遗弃。只有在少数情况下，才会在永久基地上继续投入使用。
摩尔多瓦是第一次参加这种大型活动，即使其展馆面积只有350skm，但是参与的开销也非常之大。并且2015年摩尔多瓦馆将确保专注于"滋养地球，生命能源"这一主题，在同时还需保证世博会后展馆的处置和使用。这就是世博会结束后展馆将被带回摩尔多瓦的原因。带回摩尔多瓦后，该展馆将用作当代艺术和建筑的中心和展览大厅。
通过这种方式，摩尔多瓦共和国展馆 ——"闪耀之光"（Shine the Light）—— 将展示一个关于纯能源和生态食品的特殊和多层次故事，展示可持续结构和材料的重要性。

The theme of the next world expo is "Feeding the Planet, Energy for Life" which implies that our Planet is our Home, and people on this planet have same right under the sun and each of us must have equal access to quality food, clean water, and healthy life style through good nutrition and safe energy.

Exploring the idea of pure energy internationally renowned artist Pavel Braila has proposed that the entire pavilion of Republic of Moldova to be transformed into an impressive environmental installation. On the base of Gorgona's winning project for the architectural style of the pavilion, Braila created an ambitious site specific solar installation which makes the Pavilion of Moldova to work inside and outside in the same time and that the audience begins to experience the Pavilion before it is even inside of it.

"Solar Flower" – the key element of the Pavilion, represents itself a triple mirror cube which surface consists of thousands of the facets is installed on a top of sloped metal stalk and comes just over the top of the transparent glass cube. Amirror system installed on the each corner of the glass cube follows the movement of sun all day long and reflect the sunbeam upon a rotating mirror cube creating an effect ofmyriad spots of light spinning around all over the space. Taking advantage of the transparency of the pavilion architecture, the magic effect will be observed on the walls inside and far outside of the building – on passages, terraces, lawns etc thus making every passing pedestrian being a visitor of the pavilion. The mirror installation will be also equipped with the solar batteries so the energy will be saved for the nighttimes hence making the pavilion working 24/24.

The transparency of the building and thrilling sun-effect that is obtained from the Sun – the biggest world's energy source at no costs, invites audience and clearly talks aboutglobal social concerns,environmentalheritage,openness, and equality. The sunbeam reflections all over the space besides spectacular scenery make a paradigm shift:it's time to rethink the use of conventional energies and concentrate on new pure energy with no environmental harm, preserving the natural resources and demonopolisation of the power-energy. It's also a strong mythological reference to Prometheus – who stole the fire power from gods and gives it to human, bringing enlightenment and intelligence to all the people.

The long corridor sheltered with real green plants that goes through transparent glass cube reflects attention of the architects about the environment, nature, human beings and invites audience into a bio-zone of Moldovan farming and agriculture traditions, ecological food and its famous wine which are not yet well known in the western world and to encourage visitors to get in line and be curious about what the pavilion has to tell more. Coming inside the first thing what audience encounters is a special planetarium – a series of twelve constellation made of neon on black plastic which according to Pavel Braila "these constellations are not knows by the world and can only be seen from Moldova". The constellation of Bride, Plowman, Grape, Cheese, Stork, Jaws ..First people who navigated the oceans were using stars to guide them and to discover new lands, here the stars are guiding the public through to discover the country which some people from the audience never heard of.

Artist states: "As the night progresses, the net of stars covers the sky, looking on them I'm getting inspired to tell you my story, in the way I want, much too far from the reality and much closer to another truth…"

Energy of People - Energy of Life
Entering in the main hall audience will discover the large scale projection of the film "JOC", without dialogue and without narration. Specially realized for the World Expo Milano 2015 the film features the legendary folk dance band JOC, the oldest and most legendary in Moldova. It was founded in 1945, during its long history JOC had more than 7000 shows all over the globe. The film is an ode to Moldova's authentic dance and beautiful people, great music and national costumes. Visual narration developed by Pavel Braila in this film with overwhelming details and fascinating movements will make audience to step into Moldovan Identity engaging the intimate and universal through the sensual and emotional language of dance. Dance and pure joy is the core element of Moldovan people and the embodiment of great spirit and energy for Life.

Feeding the Planet
Walking out of the exhibition hall visitors will pass by a food corner which welcomes with a nice ambiance where everybody will experience snacks and beverages made from biologically pure products, not modified genetically and grown in nature friendly conditions. Having the opportunity to try Moldovan traditional food visitors will be introduced how the products from the bar were cultivated and how agriculture works in our country by dealing with it in a new entertaining way and will show that while Organic sustainable food and cooking should be the future of Europe and entire world in Moldova this is the present and a daily life and we are ready to share with all the planet what we have.

Side specific effects
Every World Expo it is a huge machinery for representing the world's nations, and every nation is ready to put in it a lot of money Expo in Milan would not be an exception in this sense but probably because of it theme "Feed The Planet - Energy for life" this expo will make all the participants to think differently on the aspect of expenses, pollution and material waste before and after the expo. The pavilions are only contemporary structures and usually after the show practically all the pavilion a destroyed or abandoned and only in few cases they remain used on the permanent base..

Moldova is the first timer in such a huge event and participating in it is a huge expense even its pavilion area is only 350 skm and because of this Pavilion of Moldova 2015 will make sure to phocus on the idea of the them "Feed the Planet Energy for Life" but in the same time to guaranty that the pavilion will be used and after the show. That's why after the end of the show the Pavilion will be brought back to Moldova and to be used afterwards as a Center and Exhibition Hall for Contemporary Art and Architecture.

In this way "Shine the Light" the Pavilion of Republic of Moldova will showcase an exceptional and multilayered story on pure energy and ecological food alongside with the importance of structure and material being sustainable.

NOMINEE FOR BEST DESIGN AWARD OF PUBLIC ARCHITECTURE
最佳公共建筑设计奖提名奖

SoNoArhitekti（斯洛文尼亚）

获奖项目/Winning Project

米兰世博会斯洛文尼亚馆
Therma Spa by Vidalta

设计说明/ Design Illustration

森林的浩瀚、湖泊溪流的汇聚、大海和海岸线的魅力，丰富的自然保护区、品种丰富的植物群和动物群：2015年米兰世博会期间，斯洛文尼亚展馆反映了其美丽的自然风光，给人一种非常积极健康生活的感觉。

其展馆占地面积1,910m²，体现了健康环境与当地健康食品之间的紧密联系。这种联系的达成归功于一种不拖累生态系统的方法。的确，自然资源是可持续食物和良好生活质量的关键。

斯洛文尼亚展馆的目的在于刺激参观者的感官，展示其烹饪的多样性。虽然从地理角度来看，斯洛文尼亚是一个比较小的国家，但是却有24个美食区。在2015年米兰世博会，参观者可品尝大量美味的食物，包括葡萄酒，极大地满足口腹之欲。葡萄园确实是斯洛文尼亚的一个显著特征。绿地是斯洛文尼亚展馆的一个主要特点。参观者在绿地中穿行而过，会想起该国郁郁葱葱的景观，并愿意借此机会了解斯洛文尼亚所有的光彩荣耀。

展馆的设计和建造使用的是天然材料——主要为木材和玻璃。因为斯洛文尼亚是欧洲森林覆盖面最大的国家之一，所以木材可以说是该国的战略材料。展馆结构物将在内部和外部立面辅饰以内部绿墙。设计复杂，由一个大型以木材为主的组合物构成。区别于其他的展馆，该建筑结构是由斯洛文尼亚设计师设计的首个混合结构之一；它结合了木骨架结构和跨叠层木材承重构件。西立面主要是大三角形玻璃窗户，让外面的参观者可直接看到展馆内部的空间。其他立面则不如西立面般开放——主要由木板条构成。这些木板条由白色塑钢型材随机分割，看起来就像是叶子的表面。

The vastness of its forest, the abundance of its lakes and streams, the charm of the sea and its coastlines, the wealth of its nature reserves, its variety of flora and fauna:during Expo Milano 2015, Slovenia looks to reflect the image of its outstanding natural beauty, ideal for an active and healthy life. Its pavilion, which covers an exhibition area of 1,910 square meters, exemplifies the strong connection between a healthy environment and a healthy food produced locally, with methods that do not weigh on the ecosystem. Indeed, natural resources are the key to sustainable food and good quality of life.

Slovenia sets out to stimulate the visitor's senses, offering a showcase of its culinary diversity. Although, in geographical terms, it is relatively small, the country has 24 gastronomic regions, and at Expo Milano 2015 visitors can sample wide array of delights for the palate, including wines. Vineyards are indeed a characteristic of a significant portion of the Slovenian territory. Walking through the green areas, which are a key feature of the pavilion, visitors are reminded of its lush landscape and can take this opportunity to get to know Slovenia in all its glory.

The pavilion is designed and will be built with natural materials – mainly from wood and glass, as Slovenia is one of the most forested countries in Europe, we cam say thet wood is its strategic material. The structure will be complemented with greened walls inside and on the outside facades. The design is complex and consists of a large, mostly wooden composition. Distinguished by large ranges the strucutre is one of the first hybrid constructions by sloveniandesigners; it combines wooden skeleton construction with and cross-laminated timber load-bearing elements. West facade is mostly glazed with large triangular shaped windows that allow views into the interior of the exhibition spaces. Other facades are not so open – it consists of wooden battens, which are randomly divided by white steel profiles in

NOMINEE FOR BEST DESIGN AWARD OF PUBLIC ARCHITECTURE
最佳公共建筑设计奖提名奖

上海兴田建筑工程设计事务所
（中国上海）

获奖项目/Winning Project

深圳隐秀山居酒店及多功能厅
Yinxiushan Hotel Function Hall

设计说明/ Design Illustration

与自然共成长的建筑
酒店位于深圳龙岗区正中高尔夫球场，场地拥有2200亩的湖光山色。起伏延绵的山丘环绕湖face，地上植被丰富。设计充分利用地形，遵循原本的山形水势，对可保留的树木均保护利用，将建筑以环境第一的姿态植入到自然中，使其与自然交融，保留原有自然环境的状态。设计通过内外空间的联系和过渡，在首层设置多处大面积的平台与室外绿化相连接，退层的平台设置屋顶绿化，弱化了人工的痕迹，建筑好似从自然泥土中生长出来。

空间的策略
建筑中"纵、横"空间变换、交错，向自然面伸展，受到地界限制后，通过两个自然转折，将纵向空间串联在横轴上，同时缩短了300m长的单调冗长的横向动线轴。在横轴交通空间上移动和延伸过程中，纵向轴不同内容的公共空间有节奏呈现，把自然直接引入廊道，并向外部自然全部开敞，借鉴传统东方园林的廊道表达意象，自然在忽隐忽现中变得有趣、悠然。
建筑、室内、环境三位一体的设计，使建筑空间的"洗炼"、最简化成为可能。如客房通过大空间转化角度成为4.5m开间的空间，利用墙柱合一的构造体，让空间更完整，由于转角带来的唯一的三角形空间，也正好是各设备管井的最佳选择。室内空间沿用自然第一的创作原则，大量采用朴实、自然的肌理涂料，通过不同工艺产生更多肌理，通过灯光或自然光营造出不同质感的空间氛围。客房中，不规则的阳台上放置浴缸、休闲椅，顶上吊扇轻转，泥土色的墙，原木地面让人放松，在惬意中感受大自然。

"土、木"
中国人的自然观是让身心融在大自然中。建筑也不例外，它们源于自然，通过运用当代技术的提升，又恰如其分地回归到自然中，自然而然、浑然天成。在隐秀山居酒店设计中，入口雨篷高15m、跨度27m，采用木构桁架形式，具有传统工艺味道，又不失时代气息。土、木是中国人取自自然的财富，木结构的应用，体现了更多的低碳概念自然界的杉木林在其生命期内有序采伐被使用，使其可持续循环发展；木材在生产、运输、建造到解体的全建筑生命期内均比其他建材更为低碳，就是在木构建筑的使用期内，还在持续吸附着CO_2。木构建筑的防腐、防火、防裂等不利因素，通过现代技术工艺措施可大大改善，特别是构造方式，通过钢结节点混合植入使其强度大大提高，实现了建造大跨度空间的飞跃。
酒店外墙的80%均采用了涂料，这个一向被视为"低廉"的材料。施工中通过黏稠度和厚度的调整，涂料上用金属梳整出近似天然的横向肌理，取得了手工艺的设计质感；色彩选用与大地泥土同色系的基调，给人好似建筑从自然中生长出来的感觉，暗合循环、相生、互换与长久的自然观，让土、木回归。这也与当代的低碳生态观念不谋而合，一脉相承。

Architecture Coexists with Natural Environment
Located at Genzhong Golf Club of Shenzhen in Longgangdistrict, the hotel has an area of 1,466,666.67 square meters landscape of lakes and mountains. The meandering hills covered with lush forests windaround the lake. The design takes advantage ofthe original terrain and topography, protects and utilizes the reserved trees and gives priority to the concept of environment protection.Integrate the hotel with nature whilst keep the natural environment in its aboriginal status. Through the connection and transition between the exterior and interior spaces, abundant of large platforms on first floorcan access the outside landscaping, plants are set in the terraces on the withdraw layer to hide the artificial trace and decorate the building with more picturesque landscape sceneries.

Space Strategies
Horizontal and verticalspaces configurein the hotel, expand towards nature then turn around on imposed restrictions, leave the vertical spaces locate on the horizontal axis, in the meantime, shorten 300 meters of horizontal axis.While moving and extending along the horizontal direction, public spaces along the vertical direction display different contents and functions with pace, implant nature to corridor and spread outside as if invading the nature, adapted from traditional oriental garden's corridor to express image beauty, flickering in a mysterious way.
The design integrates architecture, interior environment with circumstance, makes the building a concise simplified space.By transforming space angle of guest room generated a 4.5 meters wide space, combining with wall column structure, the interior space become more complete, the unique triangle space around the corner turns out to be a perfect equipment room.Abide by the principle of giving priority to natural environment while furnishing the interior space, a great number of simple and natural texture painting were applied,different techniques formed

diversified textures and created a collection of subtle ambiance. Bathtub and leisure chair are equipped in the irregular balcony, fan turning around on the ceiling, along with the clay-colored walls and pure wood floor, the whole relaxingenvironmentgrants you a moment to feel the nature in peace and contentment.

Soil, Wood
The Chinese view of nature is to be physically and mentally in harmony with the nature.Architecture is no exception; it is originated from nature, improved by modern techniques and subtly returns to nature. While designing the Hidden Mountain Hotel, the rain cover at entrance is 15 meters high 27 meters wide, adopted a wood compose girder, traditional and contemporary.Soil and wood are wealth derived from nature, the application of wood structure expresses the concept of low carbon – to deforest and use the Chinese fir forest pursuant to its natural life cycle, keep it in a sustainable cycle; in the process of production, transportation, construction and disintegration,timbers emit less carbon than other materials, and even keep absorbing CO_2 during the period of construction.As to the adverse factors, such as anti-corrosion, fire prevention, anti-cracking, etc. can be improved by means of contemporary techniques, the construction method in particular, the strength can be improved by mixing steel node, which in the mean timerealizes the leap in construction of large span space.

80% of the exterior wall applied painting, which has always been regarded ascheapmaterial.After adjusting theviscosity and thickness, use metal to form natural horizontal skin texture; choose the color tune of soil to enhance the harmoniousrelation between the building and natural landscape sceneries, to echo the natureview of circulation, intergeneration, interchange and permanent, and to bring back soil and wood.This coincides naturally with the concept of low carbon.

NOMINEE FOR BEST DESIGN AWARD OF PUBLIC ARCHITECTURE

最佳公共建筑设计奖提名奖

冯果川 & Laura belevica（中国深圳）

获奖项目/Winning Project

南宁市规划展示馆
Nanning Planning Exhibition Hall

设计说明/ Design Illustration

项目是一座城市规划展览馆，虽然说也是公共建筑，但是公共性并不强，没有多少市民会光顾里面的展厅。这座建筑位于一座山体公园的边缘，如果按照常规做法，将会在路边留出空旷的纪念性广场，将建筑后退挤到山边，削去大部分山体建造十几米高的挡墙，这样一来既占据和破坏了市民散步的公园空间，同时也创造了一个消极的体量庞大的建筑和一个无用的广场。设计师则是将建筑贴向道路布置，以便尽可能保留山体。沿街的建筑体量托起架空形成有遮荫的近人尺度公共空间代替空旷的前广场，这里既是室外临时展厅也是供市民遛弯的通道，如此可以让城市规划展览更贴近市民。设计最显著的特征是将该建筑屋面设计为一个起伏的人造山丘，使建筑与自然山体融为一体。这不但使山体的保留变成了该项目的景观特色，还实际上扩大了山体公园的面积。这样一来虽然占据了场地、修建了建筑，但是却还给市民一个更大更有趣的开放公园。

起伏的地景屋面由若干喇叭状钢结构体支撑着，喇叭形既顺应山丘起伏又形成良好的受力形态，使建筑内部获得33m的大跨和15m的悬挑空间。喇叭同时也是导入阳光、收集雨水的装置，并且内部作为楼梯间设备间等用途。

一座规划展示馆的建筑本身也是城市建设者的宣言，这种从市民需求和环境品质出发的设计可以展示出城市管理者的以人为本的价值观。一座融化在风景中的建筑让人们看到未来城市、建筑与自然环境以及市民生活之间和谐交融的可能。设计师通过建筑设计手段调整城市的空间利益分配，实现了政府意志和市民生活的共赢。

这座建筑投入使用后，建筑周边以及地景屋面也与山体公园之间建起了围栏。设计上的公共性的设想被围栏禁锢。

The project is an urban planning exhibition hall. Though it is a public building, its publicness is not strong and few citizens come here. Since the building is located at the edge of a mountain park, according to usual practice, an open memorial plaza will be set apart by the road and most of the mountain will be razed out for building a barricade of a dozen of meters by the hillside. In this way, the park space of citizens is occupied and damaged and a negative large building and a useless plaza are created. Our approach is to have the building laid out by the road, in a view to retain the mountain to the largest extent. The building volume along the street is built on stilts to form a shady public space with an approachable scale in place of the said open plaza. Here is not only the outdoor exhibition hall but also the passageway for citizens, making the urban planning exhibition close to residents. The design is mainly characterized in that the building roof is designed to be a rolling artificial hill for integration of the building and the natural mountain. This not only makes the retained mountain become a landscape of the project, but also enlarges the area of the mountain park in fact. In this way, though we occupy the site and construct the building, a larger and more interesting open park is presented to citizens.

The rolling landscaped roof is supported by several horn-shape steel structures. The horn shape well fits the rolling hill and forms a good stress pattern, making the building have a large inner span of 33m and a cantilever space of 15m. The horns are also the devices for inputting sunlight and collecting rainwater, as well as serving as inner staircase and equipment room.

The building of a planning exhibition hall itself is also the manifesto of unban constructors. The design based on such citizen need and environmental quality can demonstrate the value concept of urban managers—people oriented. A building matching the surrounding landscape makes people have a glance at the future city, buildings and natural environment and the possible harmonious life of citizens. We aim at adjusting urban space profit distribution to realize win-win of government will and citizen life via building design.

After the building is put into use, the surrounding area and the landscaped roof are fenced with the mountain park. The publicness assumption in design is confined by fences.

409
IDEA-TOPS
艾特奖

411
IDEA-TOPS
艾特奖

艾特奖

最佳绿色建筑设计奖
BEST DESIGN AWARD OF GREEN ARCHITECTURE

IDEA-TOPS

INTERNATIONAL SPACE DESIGN AWARD

获奖者/ The Winners
Breathe.austria Prof. Klaus K.
（奥地利）

获奖项目/Winning Project
米兰世博会奥地利馆/ breathe.austria EXPO 2015 Milan

414

IDEA-TOPS
艾特奖

获奖项目/Winning Project

米兰世博会奥地利馆
Breathe.Austria EXPO 2015 Milan

设计说明/ Design Illustration

建筑设计作为景观的建筑结构

米兰世博会上的奥地利馆是一个结合建筑和环境的展示项目。通过大规模种植560m²的森林，"呼吸"创建了由人类、环境和气候组成的复杂网络。

处于内部的外部空间

展馆围绕繁茂的植物体组建框架，并充当内部景观活动的容器。凭借技术援助，框架形状产生一个奥地利森林的小气候。建筑结构中的任何光线照射处都会发生生态生长和代谢。森林面积的植被叶表面积大约43.200m²，其每小时产生62.5kg氧气，足以满足1,800人的需求，从而促进全球的氧气生产。通过蒸发冷却技术支持整个过程，而且完全不使用空调。

按照这种方式，使用相对天然的措施，即基于植物和森林土壤的蒸散冷却效应，可以重建稠密的奥地利森林。实现的气候结果与在米兰遇到的空气和气候截然不同，在展馆中可以逐渐察觉到这一点。

展示项目

展馆展示了观看技术和作为整体画面的自然环境，这可以激发许多其他项目。奥地利馆创造了一个衔接表面看来属于不协调元素的场地：技术与自然多样性百分之百的森林植被种植相对城市管理起到示范性作用，而作为景观不可或缺的一部分，它也能够为城市中的各种生命提供足够的氧气及冷风。本例凸显了奥地利的可持续绿化政策，但与此相反的是，世界范围内维持生命的树木数量正在减少。作为蓝本的该展馆可作为一种媒介，鼓励关于可再生能源、智能城市、摇篮到摇篮循环经济、零碳及绿色科技的演讲及理念。作为"空气生成站"的该展馆能够发挥气候稳定器的作用，并且能够将其功能性地集成到全世界不同规模的都市区内。对于遭受严重空气污染的城市，这种新鲜空气站能够在城市各个角落形成空气的绿洲。奥地利表明，自然与技术的混合系统能够在生态学上取得成功。

该森林包括十二种奥地利森林生态品种，带来独特的大气体验。超过190种不同物种的植物所具有的生物多样性，创造光、空气以及清新气味之间的强烈融合作用。通过雾化及洁净水喷射的辅助技术营造一种独特的小气候，为游客带来强烈的感官体验。

自5月初米兰世博会开幕起，来到奥地利展馆的游客都能够完全沉浸在清新的空气中，并且能够全身心地体验森林的美妙作用。超过12,800株多年生植物、草本植物以及森林树木已经在这个异常复杂的工程景观中安家三个多月。最先到来的白色春天使者来自于树叶之下，由蒲公英为其传播种子，清新的苔藓及云杉气息弥漫在空气中……新陈代谢——从森林开始，吸气！

The architecture. Building structure as Landscape

The Austria pavilion in Milan is a showcase project which combines building and the environment. Through the large-scale planting of 560m2 of forest, 'breathe' creates a complex network of people, the environment and climate.

The exterior space in the interior

The pavilion forms a frame around a generous vegetation body and acts as a vessel for the performance of the internal landscape. With technical assistance the framed shape produces the microclimate of an Austrian forest. Wherever light enters the built structure, ecological growth and metabolism takes place. The vegetation of the forest area has a leaf surface area of about 43.200m2 which generates 62.5kg of oxygen per hour, enough to meet the demand for 1,800 people and thus contributing to global oxygen production. This process is technically supported by evaporative cooling but is entirely free of air conditioners. In this way a dense Austrian forest can be recreated with comparativly natural measures, that is based on the cooling effect of evapo-transpiration of plants and the forest soil. The achieved climatic result differs significantly from the entcountered air and climate in Milan and becomes perceptable.

Showcase Project

The pavilion represents viewing technology and natural environments as a whole picture, that could inspire numerous other projects. The Austrian Pavilion creates a place which connects the seemingly incompatible; technology and natural diversity. The 100 percent planting of forest vegetation is an exemplary contribution to urban conduct, as the integral use of landscape can provide urban forms of life with enough oxygen and cooled air. This example highlights Austria's sustainable afforestation policy, but also its reverse in the worldwide decline in the number of life-giving trees. The pavilion as a prototype acts as an intermediary which encourages discourse and ideas on renewable energy, smart cities, Cradle 2 Cradle, Zero-Carbon and Green-Tech. The pavilion as 'air generating station' can act as a climate stabiliser and can be functionally integrated into urban areas at different scales worldwide. In cities suffering from bad air pollution, such fresh air spaces could form oases throughout the city. Austria demonstrates that hybrid systems of nature and technology can be ecologically successful.

The forest, consisting of twelve Austrian forest ecotypes, forms a unique atmospheric experience. The biodiversity of plants with more than 190 different species creates an intense interplay of light, air and fresh scents. Through technical assistance with fog and fine water jets a unique microclimate is created, creating an intense sensual experience for visitors.

From the opening of the Milan Expo in early May onwards the visitors to the Austrian Pavilion can immerse themselves in the fresh atmosphere and can experience the performance of the forest with all their senses. More than 12,800 perennials, grasses and forest trees have been installed over three months in the highly complex engineered landscape. The first white spring ambassadors come out from under the leaves and the dandelions spread their seed. The smell of fresh moss and spruce floats in the air ⋯ the metabolism – the forest begins. Breathe in!

418

IDEA-TOPS
艾特奖

获奖评语

该项目既非形象工具亦非技术工具,它提出了对大自然的敏感体验。
This Project is nor an image nor a Technical Device. It proposes a sensitive experience of Nature

NOMINEE FOR BEST DESIGN AWARD OF GREEN ARCHITECTURE

最佳绿色建筑设计奖提名奖

CHYBIK& KRISTOF ASSOCIATED ARCHITECTS（捷克）

获奖项目/Winning Project

米兰世博会捷克馆
Czech pavilion EXPO 2015 Milan

设计说明/ Design Illustration

2015年米兰世博会捷克共和国馆的国际招标被一对年轻的联合建筑师CHYBIK与KRISTOF摘得。该展馆是一幢房屋，一种体验。世博会过后，它的生命也不会结束。
此概念基于展馆的暂时性及2015年世博会的主题"滋养地球，生命能源"。
展馆格言为水。展馆展示了最新的水净化纳米技术进展，同时以游泳池的形式展示传统的捷克人共和国与水的联系（例如：水疗中心等），而游泳池属于展馆的公共空间。该展馆采用模块化结构，以方便在展览后拆卸建筑物，随后将其运回捷克共和国并循环使用展馆模块。
展馆的现代建筑设计考虑到简约的民族风格——现代主义。可在展馆的底部看到一个游泳池、小型圆形剧场以及餐厅。首层是围绕中庭的展览空间和另一个餐厅，紧随其后的是开放的绿色屋顶，游客能够透过它欣赏到整个展览会的景象。
米兰夏季的气候情况也被考虑在内，捷克展馆会向容易疲劳的游客提供额外服务。游客可自行选择在展馆中的三间餐厅之一休息，或被称为"城市泳池"或"捷克共和国公众泳池"的泳池处放松。泳池为游客免费提供必需的洗浴装备，还能将其作为宣传材料。游客可获得人字拖、一件泳衣以及一条毛巾，以取代通常会被扔进垃圾桶的传统印刷材料，而赠品上也印有关于展览会及捷克共和国的信息。游客能够在其他地方以独特的方式宣传展馆。宣传物品采用智能材料制作，其中包含了捷克共和国在世界领先的纳米技术。
当展览结束时，该展馆能够拆卸成单独的模块并运回捷克共和国以用作其他用途。该展馆将不会重蹈常规展览会展馆那样的覆辙——腐烂或拆除。底层部分将继续作为带泳池的餐厅使用，例如，用于布拉格海滨。顶部两层经简单修葺后可作为幼儿园、美术馆或学生宿舍使用。

The winning project of an international tender for the Czech Republic's pavilion at the World's Fair EXPO 2015 in Milan came from a young pair CHYBIK+KRISTOF ASSOCIATED ARCHITECTS. The pavilion is a house and an experience. But its life does not end when the World's Fair is over.
The concept is based on the temporarily of the pavilion and on the theme of the World Exhibition EXPO 2015 „Feeding the Planet, Energy for Life". The motto of the pavilion is water. The pavilion presents the latest progress in nanotechnology for water purification as well as traditional Czech Republic's relation to water (e.g. spa, etc.)in the form of the swimming pool, which is part of the public space of the pavilion. The pavilion is a modular structure. The important aspect of modularity is the ability to dismantling the building after the show, move it back to the Czech Republic and reuse modules of the pavilion.
Contemporary architecture of the pavilion refers to the simplicity of the national style – modernism. A swimming pool, a small amphitheater and a restaurant can be found at the bottom part of the pavilion. On the first floor there are exhibition spaces around a central atrium and another restaurant, these are followed by an open green roof that provides visitors with a view of the whole area of the exhibition. The interior of the pavilion is being designed in collaboration with leading contemporary Czech artists.
The climatic conditions of the summer in Milan are also taken into account, the Czech pavilion offers something extra to the often tired visitors. It is up to the visitor, if he decides to rest in one of three restaurants in the pavilion or relax in the pool which is known as „urban pool" or „Czech Re:Public Pool".

The necessary equipment for the bath gets the visitor free as promotional material. Instead of traditional printed materials that often end up in the trash, the visitor gets flip-flops, a bathing suit and a towel which should have the information about the exhibition and about the Czech Republic printed on them. The visitor then can promote the pavilion elsewhere in original way. Promotional items will be made of smart materials with nanotechnology in which the Czech Republic is a world leader.
When the exhibition is closed the pavilion can be disassembled into individual modules and transported back to the Czech Republic where it could find another use. The pavilion will not necessarily fulfill the usual fate of the exhibition pavilions, meaning decay or demolition. The lower part may continue to serve as a restaurant with a swimming pool, for example on the Prague waterfront. And after simple modification the top two floors can be use as kindergarten, art gallery or student housing.

422

IDEA-TOPS
艾特奖

423
IDEA-TOPS
艾特奖

NOMINEE FOR BEST DESIGN AWARD OF GREEN ARCHITECTURE

最佳绿色建筑设计奖提名奖

David Knafo Architects（以色列）

获奖项目/Winning Project

米兰世博会以色列馆
Israel Pavilion - Expo 2015

设计说明/ Design Illustration

以色列是一个农业研究和试验的独特实验室。以色列的农学家和研究人员在全球共享他们的知识和技术成果。

以色列馆的"未来区域"，展示了以色列在许多领域中的成就。包括在岩石土地耕种、在沙漠中种植蔬菜、新的灌溉方法和种子质量改善。

展馆的主体高度是由模块化瓷砖组成的垂直领域，可用于农作物种植。所有这些瓦片都含有计算机化的水滴灌溉系统，以优化植物生长条件。

小麦、水稻和玉米是素食食品的主要来源，将在垂直田地中生长，创造由纹理、气味和颜色组成的镶嵌图案。

展馆内的两个展厅将让参观者体验虚拟的以色列之旅，并会展现以色列科学家在农业领域研制出的革新技术。

展馆根据最先进的绿色技术设备建成，包括节能、水和空气处理。在博览会结束后，将回收整个构筑物。

在世博会的历史上，展馆建筑一直是创新和出众美感的展示。以色列馆在2015年米兰世博会上展示出一种规划方法，在这种体系结构中，农业是促进可持续发展、保护自然资源、致力于未来人类社会繁荣的加速器。

The state of Israel is a unique laboratory for agriculture studies and experiments. Israeli agriculturists and researchers share their knowledge and technology achievements worldwide.

The Israeli pavilion, "Fields of Tomorrow", is a demonstration of Israel's ability in many domains. Amongst those are cultivation of rocky land, growth of vegetables in the desert, new methods of irrigation and improvement of seeds quality. The main elevation of the pavilion is a vertical field composed of modular tiles used for the cultivation of agricultural crops. Each of these tiles contains a computerized drip irrigation system to optimize the plants growth conditions. Wheat, rice and corn, which are the main sources of vegetarian food, will be grown on the vertical field creating a mosaic of textures, smells and colors.

Two exhibition halls inside the pavilion will allow visitors to take a virtual tour of Israel, and will present the technological innovations in agriculture, developed by Israeli scientists.

The pavilion has been built under most advanced green technology devices including energy saving and water and air treatment. The whole structure will be recycled at the end of the exposition.

Along Expo history, architecture of pavilions has been a demonstration of innovation and outstanding aesthetics. The Israeli pavilion in Milan 2015 presents an planning approach, in which architecture is a vehicle to promote sustainability, protection of natural resources and dedication to social prosperity for the future of mankind.

May

July

September

October

NOMINEE FOR BEST DESIGN AWARD OF GREEN ARCHITECTURE

最佳绿色建筑设计奖提名奖

Kadarik, Tüür. Arhitektid.

Kadarik, Tüür. Arhitektid（爱沙尼亚）

获奖项目/Winning Project

米兰世博会爱沙尼亚馆
Milan EXPO Estonian pavilion

设计说明/ Design Illustration

爱沙尼亚馆的标题，"画廊"，象征着整个展馆的性质，更广泛地说，象征着爱沙尼亚作为一个动态、智能的小国的理念，其公民的进取精神、引入的外商投资、国际层面的合作和外国投资者都影响着它的发展。

"画廊"为具有创意的爱沙尼亚人提供了一个平台，他们将会在其中填满生活和内涵。整个展馆的结构以它需要为承办的各种演出、展览、活动和演示创造最佳条件的理念为基础来建造而成。展馆将不仅仅为形式而生，它更是为能够赋予展馆以生命的内容提供一个框架。

展馆的理念建立在北欧民主价值观之上：尊重每个人表达自我和发挥创造力的权利。更重要的是，仿佛这确实是一个充满活力的民主小国的典范：任何人在那里做了什么或哪些人去那里访问都可能会影响它的发展情况。这将会形成集体创作的成果，并且可以通过新的理念和现代应用提供补充。爱沙尼亚富有创意的人们以及所有的参观者将造就展馆的最终性质，展馆将通过自然、科技、文化和美食的集体影响为参观者提供爱沙尼亚体验。

更广泛地说，一个开放的平台也意味着透明和灵活的营商环境，这有利于实施新举措以及成为全球进程中负责任的一部分。进行更近距离的观察后，可以发现这个想法展示了我们在爱沙尼亚引以为傲的，也同时是我们希望在展馆二楼放大的所有活动和举措，在旋转的主题展会上：将在展馆二楼提供一个单独的展览区域以用于集中体现创新型企业、电子国家解决方案、绿色乡镇企业、乡村旅游、创意产业和艺术。所有展示的主题和我们讲述的故事都将与展馆中遍布的主题交织在一起——爱沙尼亚是一个环境优美、充满创造和创新的国家。

通过展馆的绿色主题和随后的模块回收，来推广爱沙尼亚有机原材料和优秀厨师的粮食计划，而一楼和二楼的常设展览则讲述原汁原味的爱沙尼亚故事，通过使用当代技术和设计为故事赋予新的形式和功能，并将展馆与世博会的主题，"滋养地球，生命的能源"联系起来。

参观者可在展馆内自由选择举办何种活动、事实或故事编织成他们个人的爱沙尼亚体验。任何进入展馆的人都应该记住最重要的是一种温馨、开放、热情好客的氛围，这些将通过令人兴奋的事实、扣人心弦的故事、天籁之音、优雅的技术简约之美、美味的食物和动听的音乐得到充分展示。

展馆2.1的描述

创建展馆的目的是为了建立一个举办演出和展示创意成果的多功能环境。空间将主要通过在展馆中举办的活动来体现其意义和氛围（音乐会、展览、演讲、会议等）。

该展馆的结构假定是成为具有明确定位的空间，使参观者了解到爱沙尼亚的真实情况，所以将该展馆建造为自然和创造力的画廊。

布局和功能性的描述

展馆由像魔方一样离心式堆叠的"巢箱"组成，形成面对俄罗斯馆的主要建筑体积，同时也沿着这一分区内的道路。

"巢箱"是一种高顶空间，创造出类似腔室的、分隔的空间区域。世博会结束后，这些模块化的组装部件将运回爱沙尼亚并在爱沙尼亚用作儿童游戏场、自然观赏点或者，如公共汽车站候车亭。从理论上讲，可以拆卸整个爱沙尼亚馆，并在爱沙尼亚完全按照相同的样式建造起来。

"画廊"的主开口朝向南方，走过这条东西方向的画廊时，需穿过高大、耸立的雨篷，这些雨篷连同宽敞的内部形成了一种震撼的体验。此外，设置了从广场通往北方且位于街道方格之间的入口。

就其开放性而言，不仅意味着展馆既欢迎世博会参观者，也暗示着爱沙尼亚对创新和世界的开放。

该建筑将建有三个楼层。一楼设有带有配套场所、酒吧和纪念品柜的厨房。此外，形成舒适"隔间"的"巢箱"将与位于一楼的展览空间的其余部分共同为参观者提供了解有关爱沙尼亚信息的机会。

在刚进入东西大街后，可见到一个规划舞台区域，该舞台由1.2m×1.2m的模块组成，足够灵活并能够根据需要轻松进行重新布置。

沿着建筑西侧的一条狭窄楼梯和来自建筑中心的一个通风双螺线楼梯将会成为前往一楼的通道。在一楼，主展厅介绍了爱沙尼亚、的企业家和各行业，参观者可以在那里深入探索各个行业取得的成就。同样，一楼设立了一个"黑麦酒吧"讲述爱沙尼亚作为一个独特的国家生产清洁黑麦的故事。

二楼设有一个将爱沙尼亚风格的植物和树木作为主基调的屋顶花园。植物种植在专用容器中，容器之间带有间距以供行走或闲坐、吸引参观者，例如，如果他们愿意，可以使用从一楼购买的能够携带的野餐用具在这里野餐。

通过楼房北部的楼梯井进入到2楼。在小广场上方紧邻楼房后方的北部设置有休息室，规划用于为一小组人员坐在软体家具的舒适座椅上举行商务会议，同时还可以观赏2楼的景色。

此外，二楼设有展馆工作人员和管理人员的杂物室和办公室。

三个楼层都通过楼梯相互连接，主楼梯向所有人开放，后部楼梯用作逃生楼梯或工作人员在各个楼层之间工作使用。此外，楼房还为行动不方便人士安装有平台升降机。

室内建筑理念内部结构的设计基于外部和内部空间的单一综合设计，通过使用各种木制表面强调统一性，沿着楼房的像素结构创造有节奏的重复。按照尽可能多的功能设计空间，这样可以根据需要和展馆举行的项目重新排列。内部设计较为简单，是一个旨在作为中性背景以用作广泛使用的LED屏幕演示空间。

基于空间的合理性，木箱模块将采用不同的标准设计：悬挂模块、演示模块、1楼LED显示屏展示模块、2楼旋转展厅模块、会议室模块及配套楼宇模块。

空间和功能模块概念

一楼开放综合区有着种满盆栽植物的绿色柜台，创造了一个清新和灵活的空间，将可举行各种项目活动。一楼综合区域与有节奏交替的能量摆动巢和配有LED屏幕的常设展览巢交界。附属楼宇附近设有自动饮水器，以便为参观者提供洁净饮水。

一楼可提供更多的隐私空间。

在中部，设有一个黑麦主题酒吧，还有一个安装有悬挂椅以供人们闲坐的区域，这与可用于闲谈和提供食物和饮料的几排配有酒吧高脚椅的桌子形成互补。1楼的空间理念仍在这里延续，在这里，展览空间与私人能源摇摆巢相互交错。

二楼的屋顶花园向所有有兴趣的人群开放，它的旁边就是用于特邀嘉宾和会议预约的会议室。将在屋顶花园的苗木箱之间提供座位模块以及装饰有暗淡和温暖灯光设计的绿洲。屋顶花园的四周为画廊的遮阳篷，可通往办公室或会议室区域。

整栋楼的角落和缝隙以及私人巢的概念将由展馆内照明设计支持，其可以根据自然夜间或白日光线条件自动调节。

The title of the Estonian pavilion, "Gallery of _", symbolises the nature of the entire pavilion and, more broadly, the idea that Estonia is a dynamic and smart small country, with every citizen's initiative but also every foreign investment, collaboration with international reach and foreign investor affecting how it fares.

"Gallery of __" is an open platform for creative Estonian people, who will fill it with life and content. The architecture of the entire pavilion is created based on the idea that it should create the best conditions for holding various performances, exhibitions, actions and presentations. The pavilion will not be form for form's sake but rather a framework for content that will bring the building to life.

The philosophy of the pavilion rests on democratic Nordic values that respect every individual's right to self-expression and creativity. What is more, it as if this is indeed a model of a dynamic democratic small country:how it fares may be influenced by all who are doing something there or who visit there. Collective creative output will be formed, which can be complemented by new ideas and contemporary applications. The final nature of the pavilion will form up with Estonian creative people but also with all the visitors, who will be provided with an experience of Estonia through the collective impact of nature, technology, culture and cuisine.

More broadly, an open platform also means a transparent and flexible business environment that favours new initiatives, as well as responsible connection to global processes. Upon closer inspection, this idea informs all the activities and initiatives that we are proud of in Estonia and that we wish to zoom in on the 2nd floor of the pavilion, in rotating themed exhibitions:a separate exhibition area will be provided on 2nd floor of the pavilion for the focused presentation of innovative companies, e-state solutions, green rural enterprise, rural tourism, creative sectors and fine arts. All presented themes and stories told by us will be intertwined with the pervasive theme at the pavilion:Estonia as a country of nature, creativity and innovation.

The pavilion will be linked to the main theme of the EXPO, "Feeding the Planet, Energy for Life", by the green execution of the pavilion and the subsequent recycling of its modules, a food programme paying tribute to organic Estonian raw ingredients and talented chefs, and the permanent exhibition recounting on both the ground floor and 2nd floor, Estonia's stories born in nature, given a novel form and functionality by means of contemporary technology and design.

Guests at the pavilion have free choice of what activities, facts or stories to weave into their personal experience of Estonia. The most important thing that anyone entering the pavilion should remember is a warm, open and hospitable atmosphere, laced with exciting facts, gripping stories, rare sounds of nature, elegant technological simplicity, tasty bites and good music.

Description of the pavilion 2.1. Central idea

The pavilion is created to be a multi-functional environment for holding performances and presenting creative output. The space will draw its meaning and atmosphere primarily

from events held at the pavilion (concerts, exhibitions, presentations, conferences etc).
The architectural premise for the pavilion will be a space with a clear identity, enabling visitors to get an idea of what Estonia is about. The pavilion is structured as a gallery of nature and creativity.

Description of the layout and functionalities
The pavilion composes of "nestboxes" stacked off-centre like cubes, forming the main volume of the building towards the Russian pavilion and along the side of the road inside the quarter.
A "nestbox" is a high room creating a chamber-like sectioned-off area of space.
After the end of the EXPO, these modular-assembly elements would travel back to Estonia and be used there as children's play grounds, natural viewing points or, for instance, bus stop shelters. Theoretically, it will be possible to dismantle the entire Estonian pavilion and to set it up in the exactly the same form in Estonia.
The main opening of "Gallery of_" towards the south, Decumanus, entry from which, under a high, lofty awning, along with the spacious interior creates a powerful experience. Additionally, there will be entrances from the plaza to the north and between boxes from the street.

With its openness, the pavilion both invites EXPO visitors and allude to Estonians' openness to innovations and the world.

The building will have three floors. The ground floor houses a kitchen with ancillary premises, bars and a souvenir stand. Furthermore, "nestboxes", forming cosy "compartments", which together with the rest of the exhibition space on the ground floor provides visitors with an opportunity to quench their thirst for information about Estonia.
Immediately upon entering off Decumanus, there is an area planned for a stage which, consisting of 1.2 x 1.2 m modules, will be flexible and capable of being readily re-arranged as needed.
Access to the 1st floor will be provided by a narrow stairway will lead along the west side of the building and an airy double-flighted stairway from the centre of the building. On the first floor, the main exhibition area profiling Estonia, our entrepreneurs and sectors, where visitors will be able to explore in-depth sectoral achievements. Similarly, the first floor will house a "rye bar" recounting the story of Estonia as a unique country producing clean rye.
The 2nd floor houses a roof garden where plants and trees characteristic of Estonia set the tone. Plants are placed in dedicated containers with space in between for walking or sitting, inviting visitors, for instance, to also have a picnic, if they so desire, with a picnic set bought from the ground floor to take away.
Access to the 2nd floor is provided through the stair wells on the north side of the building. Adjacent to the rear northern side of the building above the small plaza, a

lounge has been planned for holding business meetings for a smaller group of people in a cosy setting on soft furniture with a view of the garden on the 2nd floor.
In addition, the second floor houses utility rooms and premises for the staff and management of the pavilion.
All three floors are inter-connected by stairs, with the main stairs open to all and the rear stairs used as escape stairs or by staff moving between various floors.
Furthermore, a platform lift will be installed in the building for mobility-challenged persons.
Interior architecture Interior architecture concept
The interior architecture design is based on a single comprehensive design for the exterior and interior space, doing away with boundaries between the two.
Comprehensiveness is emphasised by the copious use of wooden surfaces, creating rhythmic repetitions along with the pixel structure of the building. The design of the space has been created to be as multi-functional as possible, so that it may be re-arranged according to need and the programme happening at the pavilion. The interior design will be simple and intended as a neutral backdrop for the widely used LED-screen presentation space in the interior.
Based on the rationale for the space, wooden box modules will employ various standard designs: swing module, presentation modules with LED screens on floor 1, rotating exhibition module on floor 2, conference room module and blocks of ancillary premises.

Concept for spatial and functional modules
The open general area on the ground floor with a green counter of potted plants create a fresh and flexible space enabling various programme events to be held. The general area on the ground floor is bordered by rhythmically alternating energy swing nests and permanent exhibition nests with LED screens. Near the block of ancillary premises, there are drinking fountains of clean water for visitors.
First floor provides more privacy.
In the middle, there is a rye-themed bar and an area with suspended chairs for hanging out, complemented by groups of tables with bar stools for conversation and supporting food or drink. The spatial rationale of floor 1 continues, with exhibition space alternating with private energy swing nests.
The roof garden on the second floor are open to all those interested; it is bordered by conference rooms for invited guests and meeting appointments. Between plant boxes in the roof garden there will be provided sitting modules and an oasis with a dime and warm lighting design. The roof garden is surrounded by a gallery with an awning, giving access to the block of offices/conference rooms.
The concept of nooks and crannies and of private nests for the entire building will be supported also by the lighting design at the pavilion, which will be automatically adjustable according to the natural night/day light conditions.

NOMINEE FOR BEST DESIGN AWARD OF GREEN ARCHITECTURE

最佳绿色建筑设计奖提名奖

深圳毕路德建筑顾问有限公司
(中国深圳)

获奖项目/Winning Project

国电宁夏太阳能有限公司办公楼
State Elatronic Solar Energy Ltd Office Building

设计说明/ Design Illustration

国电宁夏多晶硅厂区是以多晶硅为主要产品的企业，多晶硅的需求主要来自半导体和太阳能电池，是高集成的高科技产品，把握企业形象和满足行政办公的需要是我们追求的目标。

本建筑设计从建筑室内空间需要和造型出发，结合企业产品特点，将多晶硅矿石的造型和现代化的审美情趣结合，打造出别具一格的建筑空间体验，利用太阳能节能环保企业特性，将生态建筑的概念引入建筑设计当中。

立面采用现代感强的合金镀锌波纹板，以不同的颜色和独特的造型特点作为立面设计的主要手法，在宁夏地区灰大干旱的气候条件下，防止风沙腐蚀，红白色立面主要色彩使得建筑成为该地区吸引视线的地标性建筑，金属色彩斑斓的外表皮也映衬出现代化企业的高科技兴性质。

由于宁夏地区室外条件不利于人们长久活动，通过绿色生态的设计手法将不同植物引入到建筑内部，通过温室效应，打造宜人的室内活动小环境，拓展室内空间，使室内活动场所能代替户外活动的需求。

景观：闪烁的绿洲

一改传统办公楼直达正入的方式，而是利用"绿意的折线"环绕在"钻石"周围，利用原生景物形成一个"生态绿洲"。选用当地原生植物品种：虎尾草、柴叶小檗、金叶荻、金露梅、火炬树、沙棘等，使整体景观随季节不同产生变幻。折线台地，既解决了建筑入口高差的问题，也在钻石造型前形成了一个诗意化的"托盘"。夜晚，建筑及景观在灯光的照射下，闪现出色彩斑斓的颜色，成为地域内最醒目的视觉焦点。

室内：澄净的湖水

接待厅将人们带入一个四季皆宜的户外环境，每层员工都可以享受这一空间。巨大的接待背景好似贺兰山的一角，宽敞的空间、光滑的地面形似山间的盐湖，水面上散落着贺兰山的山石。木饰面、玻璃隔断、地毯、石材等组成了和谐的自然环境。自然光线充足、内部空气流通顺畅，视觉体验灵活，为员工提供了舒适，高效的办公环境。钢的本色随处可见，钢柱脱离了规规矩矩的组织方式，在工业化的粗犷之上有一种自然的灵活和温馨。

Guodian Ningxia Polycrystalline Silicon Factory is specialized in producing polycrystalline silicon that is a highly-integrated high-tech product widely used in semiconductor and solar cell. Our ambition is to grasp the concept of enterprise image and meet the needs of executive office.

According to the interior space demand and modeling of the building, the design combines the product characteristics of the enterprise and the modeling of polycrystalline silicon ore with modern aesthetic taste, in order to develop a unique building space experience and introduce the concept of ecological construction into building design by virtue of the features of the solar energy enterprise.

The façade adopt alloy galvanized corrugated plate with different colors and unique modeling characteristics, which enhances its modern sense. Under the dry and dusty weather conditions of Ningxia, the façade can prevent sand corrosion. It is designed in red to attract sight as a landmark. The colorful metal surface suggests the high-tech property of the modern enterprise.

As the outdoor conditions of Ningxia are not favorable for people's long-termactivities, the design of green ecology introduces different plants into the building and creates a small indoor activity environment via greenhouse effect, which expands the indoor space and makes the indoor activity place replace people's outdoor activity demands.

Landscape: twinkling oasis

An "ecological oasis" is formed by taking advantage of the native landscape by surrounding the "diamond" with the "green broken line", instead of thetraditional direct and front entry of office buildings. The use of the native plants, such as Chloris virgata Sw, Berberis thunbergii DC.var. atropurpurea Chenault, caryopteris clandonensis, Potentilla fruticosa, Rhus Typhina and Hippophae rhamnoides Linn, makes the landscape change in different seasons. The broken line terrace meets the entry requirements of people of different heights and forms a poetic "tray" in front of the diamond modeling. At night, the building lightens out gorgeous colors under light irradiation, becoming the most striking local visual focus.

Indoor: clean and clear lake water

The reception hall brings in a standard outdoor environment of four seasons for people of this office. All floors have a platform to share this space. The huge reception backdrop is like a corner of Helan Mountain. The inner space is capacious. The smooth floor

is like an intermountain salt lake dotted with hills tones of Helan Mountain. A harmonious environment is formed by wood finish, glass partition, carpet and stone. With abundant natural light, smooth air circulation and flexible visual experience, the building provides employees a comfortable and efficient office environment. The true quality of steel is everywhere. Steel columns are organized in a natural, flexible and cozy way.

435
IDEA-TOPS
艾特奖

获奖者/The Winners
michele molè
（意大利）

获奖项目/Winning Project
米兰世博会意大利馆
Italy Pavilion Expo 2015

获奖项目/Winning Project

米兰世博会意大利馆
Italy Pavilion Expo 2015

设计说明/ Design Illustration

意大利展馆设计

在2013年5月的国际竞赛中，Expo 2015 S.p.A. 将意大利展馆设计裁定为胜出项目。共有68件建筑作品参与角逐，优胜设计来自于与Proger合作的Nemesi及Proger和BMS Progetti（从事工程及费用管理）以及Livio De Santoli教授（从事建筑物可持续性）。

"我们设想的建筑将可代表共处的理念，以及通过创新的当代建筑物，并考虑到伟大的意大利建筑传统，将自身视作社区一份子的能力。"——Michele Molè，Nemesi的创立者及董事。

意大利展馆由创建的意大利宫殿（建筑面积14,400m²，六层）及Cardo旁的一些临时建筑（建筑面积12,500m²，二层）组成。
从初步设计到执行设计耗时7个月；
超过200名专家参与为期14个月的施工；
由意大利公司进行展馆建设；
绝大部分的展览空间，共计27,000m²

意大利展馆建筑设计

凭借其建筑设计及坐落于四个方位基点中的一个偏北基点上，意大利宫殿显得极为突出，成为世博会场上的一个真正地标。同时，它也为现场右侧的Viale del Cardo提供戏剧背景。
意大利宫殿高度为35m，是世博会场的最高点。它是世博会唯一的永久存在建筑物。
意大利宫殿利用"城市森林"的概念，采用Nemesi设计的"枝状"外壳，以便同时创造出原始及科技图像。线条的交织带来光、影、实体与空隙的交相辉映，产生带有清晰陆地艺术的雕塑般建筑。
对于Nemesi而言，意大利宫殿的闪光是一种聚合概念，其中的吸引力产生一种重新认识社区及其附属物的感觉。内部广场代表了社区的能量。这一空间——综合建筑物的标志性核心——是展览会路线的起点，处于组成意大利宫殿的四个建筑体的中间。这四个建筑体容纳有：会展区（西）、礼堂及活动区（南）、办公区（北）和大会与会议区（东）。这些建筑体是巨型树木的象征，采用模拟嵌入泥土的大型树根的巨大底座。从广场内部来看，它们都保持开放，当向上看时视线变长，看起来像是形成了巨大光滑屋顶上方的天篷。

意大利宫殿的功能布局

Nemesi将这个意大利宫殿的展览布局设计为循序渐进的过程，以便慢慢发现并领会这个特别建筑景观的外形与内涵。该路线始起于内部广场，这是一个接待游客的大厅。弯曲倾斜的立面，为这一建筑体带来流动及活力感，形成纯正美丽的空间。盘旋而上的阶梯从广场处蜿蜒上升，纵向穿过这一区域，像是连接了所有楼层。

创新及可持续办法

意大利宫殿是当代工厂的象征，意大利建筑及建设上的一次挑战，是保留用设计、材料及技术方面的实验性与创新性为标志的一项工程。意大利宫殿设计并构思为一座可持续能源建筑，而由于表面覆盖的光伏玻璃及外壳新型混凝土的光媒作用特征，其能源消耗几乎为零。与环境沟通并交换能源的"渗透性"建筑。
2,000吨不活跃的生物动力混凝土覆盖700种各式各样的树枝状板材 4,000m²翼板覆盖——400吨钢材形成外壳的分枝之间的充分交织，凸显出意大利宫殿的雕塑般外观。Nemesi使用独特的几何学设计，以创造出这层外部"皮肤"。意大利宫殿正面的全部9,000m²表面覆盖着超过700块不活跃的生物动力混凝土板，该板材采用了Italcementi取得专利的TX激活技术。当材料与光线接触后，能够"捕捉"空气中污染物，转化为惰性盐并降低雾霾水平。砂浆采用80%的再生骨料，包括Carrara大理石采矿场的废弃材料，有助于在传统白水泥中增强光泽。这种新型材料同样也是非常"动态的"，从而创造出在此板材中应用的流体设计，这属于意大利宫殿施工的一个环节。
所有外壳的板材都是通过Styl-Comp创造的独一无二的部件。
Nemesi为意大利宫殿设计的屋顶是由Stahlbau Pichler创造的一种创新型"翼板"。它通过光伏玻璃、平面及曲面的几何外观（通常为方形）诠释森林苍穹。再加上该建筑的"树枝状"外壳，这些都是设计与技术的创新表现。屋顶达到内部广场以上的建筑高度，巨大的光滑锥形天窗"横亘"在广场及中间阶梯上，散发着自然光。

沿着CARDO的建筑物

意大利展馆包括可从Cardo向外看到的一系列临时建筑物。这些建筑都采用"干式"技术建造而成，以便在世博会结束时进行拆卸及重新配置。
Nemesi采用意大利乡村思想沿着Cardo开发该建筑物，创造出与小型广场、露台及人行道并列的建筑体。将不时向外突出的不同几何形状组合，几乎形成巨大的镶嵌图案，而每一小块都有自有图样和自主性。地面与首层通常是由交错的建筑体形成的小广场。
沿着Cardo的建筑物代表着意大利的不同组成部分，尤其是提供会议及展览区的区域。沿着Cardo北部、面向意大利宫殿的区域用于分属欧盟展馆的机构、展会及会谈，象征着欧洲与意大利之间的紧密关系。

DESIGN OF THE ITALY PAVILION

The design chosen for the Italy Pavilion was the winning project in an international competition adjudicated by Expo 2015 S.p.A. in May 2013. In total, 68 architectural practices took part, the winning design came from Nemesi&Partners in association with Proger and BMS Progetti (for the engineering and cost management) and Prof. Livio De Santoli (for building sustainability).

"We imagined an architecture that would represent the idea of being together and the ability to recognize themselves as a community through an innovative, contemporary building taking into account the great tradition of the Italian architecture."Michele Molè, founder and director of Nemesi.

The Italy Pavilion consists of the creation of Palazzo Italia (built area 14,400m2 with 6 levels) and some temporary buildings along the Cardo (built area 12,500m2 with 2 levels).

7 months from the preliminary to the executive design
14 months of constructionover 200 professionals involved
Italian companies to build the pavilion
27,000 square meters in total, for the most part exhibition spaces

ARCHITECTURE OF THE ITALY PAVILION

Palazzo Italia is a genuine landmark on the Expo site, standing out because of its architecture and its location on one of the four cardinal points – the northern one. It also provides a scenic backdrop for Viale del Cardo which runs right across the site.
Palazzo Italia reaches a height of 35 meters, the highest peak within the Expo site. It's the only permanently architecture at the Expo.
Palazzo Italia draws on the concept of an "urban forest", with

the "branched" outer envelope designed by Nemesi to simultaneously conjure up primitive and technological images. The weave of lines creates a play of light, shadow, solids and voids that generates a sculpture-like building with clear hints of land art.

For Nemesi, the spark for Palazzo Italia was a concept of cohesion in which the force of attraction generates a rediscovered sense of community and belonging. The internal piazza represents the community's energy. This space – the symbolic heart of the complex – is the starting point for the exhibition route, in the midst of the four volumes that make up Palazzo Italia. These four volumes house the Exhibition zone (West), the Auditorium and Events zone (South), the Office zone (North) and the Conference and Meeting zone (East). The volumes are symbols of giant trees, with massive bases that simulate great roots plunging into the earth. Seen from the internal piazza, they open up and become longer as you look up, visually forming a canopy beyond the giant glazed roof.

FUNCTIONAL LAYOUT OF PALAZZO ITALIA
Nemesi designed the exhibition layout of Palazzo Italia to be a gradual journey to discover and understand the shapes and contents of this special architectural landscape.

The route starts from the internal piazza, a great hall in which visitors are welcomed. The curved, inclined elevations give the volumes a sense of fluidity and dynamism, forming a space of genuine beauty. The great flight of steps, that rises up from the square, crosses this area longitudinally to visually connect all the floors.

INNOVATIVE AND SUSTAINABLE APPROACH
Palazzo Italiais the symbol of the contemporary factory, an Italian architectural and constructive challenge, a work characterized by experimentation and innovation in terms of design, materials and technologies used. Palazzo Italia was designed and conceived as a sustainable energy building almost zero thanks to the contribution of photovoltaic glass in coverage and photocatalytic properties of new concrete for the outer casing. An " osmotic" building that dialogues and exchanges energy with its surroundings.

2,000 tons of i.active Biodynamic concreteover 700 branched panels all differents
4,000 sqm of sail covering – 400 tons of steel
The rich weave of branches that forms the outer envelope helps to highlight the sculpted shapes of Palazzo Italia. Nemesi used a unique geometric design

to create this outer "skin". The full 9,000m2 of the façade of Palazzo Italia is clad in more than 700 i.active BIODYNAMIC concrete panels with Italcementi's patented TX Active technology. When the material comes into contact with light, it can "capture" pollution in the air, transforming it into inert salts and reducing smog levels. The mortar used is 80% recycled aggregates, including scrap material from marble quarries in Carrara that helps add more luster than in traditional white cement. This new material is also very "dynamic", enabling the creation of fluid designs like the complex shapes used for the panels that are part of the construction of Palazzo Italia. Allpanels for the envelope are unique pieces realized by Styl-Comp.

The roof designed by Nemesi for Palazzo Italia is an innovative "sail" realized by Stahlbau Pichler. It's an interpretation of a forest canopy, with photovoltaic glass and flat and curved geometric shapes (often squares). Together with the building's envelope of "branches", it's a manifest expression of innovation in design and technology. The roof reaches its architectural height above the inner piazza, where a massive glazed conical skylight "hangs" over the square and the central steps, radiating natural light.

BUILDINGS ALONG THE CARDO

The Italy Pavilion includes a series of temporary buildings that look out onto the Cardo. These are being built using "dry" technology to facility their removal and relocation when the Expo ends.

Nemesi used the idea of an Italian village to develop the buildings along the Cardo, creating volumes juxtaposed with small squares, terraces and covered walkways. Different geometric shapes, at times overhanging, are assembled to almost form a giant mosaic in which each piece has its own design and autonomy. The ground and first floors are often little squares produced by the alternation of architectonic volumes.

The buildings along the Cardo represent the different parts of Italy, especially the regions, providing them with meeting and exhibition areas. The area facing Palazzo Italia, along the northern part of the Cardo, is used for the institutions, exhibitions and meetings that fall under the European Union Pavilion, providing a symbol of the close ties between Europe and Italy.

442
IDEA-TOPS 艾特奖

获奖评语

内部公共空间宽敞,外部动人,妙用了一种抗烟雾水泥。意大利在2015世博会的永久展馆标志着BIM技术在设计和施工中的重要应用。

Rich spatially in the articulation of interior public spaces evocative on the outside , wise in the use of a anti smog cement ,the Italian permanent building at Expo 2015 represents also an important implementation of BIM Technology in design and construction

NOMINEE FOR BEST DESIGN AWARD OF DIGITAL ARCHITECTURE

最佳数字建筑设计奖提名奖

JakobMacFarlane（法国）

获奖项目/Winning Project

FRAC艺术中心
FRAC Centre - Center for Art and

设计说明/ Design Illustration

由委托人与Région Centre投资，并得到国家、欧洲及Orléans市的支持，这幢新建筑使得FRAC中心有资格成为世界唯一的建筑实验室。
Jakob与MacFarlane提出一种动态形式的突起——通过挤压现场现有建筑物的结构格架并将其扭曲。强烈的建筑信号与其环境融合，这一流动式混合构筑物包括三层玻璃和金属"突起"，而内部庭院中是具有历史价值的一组现有军事建筑。
将庭院作为公共场所，充当建筑物和住宅与FRAC新方案之间连接的一个地形表面。新的表面利用了现场的天然地形，将游客自然的引领至建筑入口。地面加强了三段紊流的视觉动力，并利用其有机几何形状通往城市。
FRAC中心通过其崭新的都市外立面，与Orléan文化城市网络连接起来，将庭院转变成城市空间。这个全新的建筑物通过新颖的结构和几何造型，成为了整个现场的质心。别出心裁的原型尺寸，与FRAC中心及其藏品相得益彰。
建筑物的LED外立面，诞生于Electronic Shadow艺术家们的合作，向城市播送FRAC的实时信息。体、面、线汇合，形成一个时刻变化着的动态交汇建筑。最后，建筑物逐渐非物质化，变成一个光影信号。

Funded by the client, Région Centre, with the support of the State, Europe and the City of Orléans, the new building allows the FRAC Centre to assert its role as the world's single laboratory for architecture. Jakob+MacFarlane proposed the emergence of a dynamic form by extruding the structural grid of the site's existing buildings and distorting them. A strong architectural signal, interacting with its context, this fluid and hybrid structure consists of three "emergences" composed of glass and metal within the interior courtyard of a group of existing historic military buildings.
The courtyard is treated as a public space, a topographic surface that serves as a link between the buildings and the home to the FRAC's new program. This new surface plays with the natural topography of the site, leading visitors naturally to the building's entrance. This surface reinforces the visual dynamics of the three turbulences and reaches towards the city through its organic geometry.
With its new urban façade, the FRAC Centre is connected to the cultural urban network of Orléans transforming the courtyard into a city space. This new architectural presence has become the center of gravity of the overall site, with a new structure and geometry. Its distinctive prototypical dimension resonates with the identity of the FRAC Centre and its collection. The building's LED façade, conceived in collaboration with the artists Electronic Shadow, displays real-time information about the FRAC to the city. Volumes, surfaces and lines merge to create a dynamic and interactive presence in flux. Ultimately, the architecture dematerializes into a signal of light.

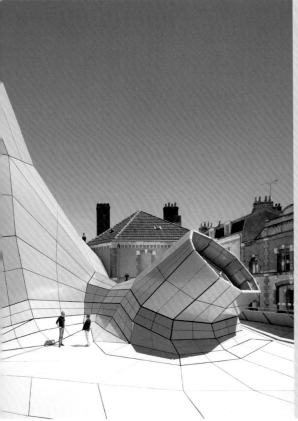

NOMINEE FOR BEST DESIGN AWARD OF DIGITAL ARCHITECTURE
最佳数字建筑设计奖提名奖

JakobMacFarlane（法国）

获奖项目/Winning Project

绿立方——欧洲新闻电视台总部
The Green Cube - EuronewsTelevi

设计说明/ Design Illustration

"绿立方"设计作为欧洲电视新闻频道——欧洲新闻电视台的全球总部，是沿着法国莱昂塞纳河畔的新建建筑，意图与塞纳河形成互动（参见航拍城市图片）。

本项目设计为拉伸的立方体，穿插着两个圆锥形中庭，方便透进阳光、空气，并为建筑使用者与公众带来河道美景。这两个中庭像是两个巨大的眼睛，注视着河面及周边环境。

这两个眼睛也代表着欧洲新闻频道的眼睛，一对抽象的接受体，不断捕捉着我们周围世界中的事件。

"绿立方"构想的竞争建筑是早在2010年在河畔建成的"橙立方"（参见城市航拍图片）。其立面及圆锥形孔洞概念灵感来自于毗邻的塞纳河——其流动、颜色、气息及环境。

The 'Green Cube', designed as the world headquarters of the television news channel Euronews, is a new building built along, and imagined in dialogue with, the Saone River in Lyon, France (see aerial urban photo).
The project is designed as an elongated cube pierced by two conical atriums introducing daylight, air and providing views to and from the river for the building's users and the public. These atriums are imagined as two gigantic eyes looking onto the river and its environment.
Symbolically, the eyes also represent those of Euronews, abstract receptors, capturing the events of the world around us.
The 'Green Cube' was conceived for the same architecture competition as the 'Orange Cube', which was completed earlier in 2010 further along the riverside (see aerial urban photo). Their façades and conic voids were conceptually inspired by the adjacent river – its movements, colors, moods and atmospheres.

449
IDEA-TOPS
艾特奖

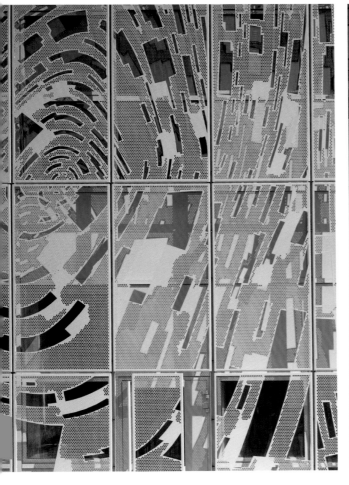

451
IDEA-TOPS
艾特奖

NOMINEE FOR BEST DESIGN AWARD OF DIGITAL ARCHITECTURE

最佳数字建筑设计奖提名奖

dream Design factory
（土耳其）

获奖项目/Winning Project

米兰世博会土耳其馆
TURKISH PAVILION EXPOMILANO2015

设计说明/ Design Illustration

设计师的建筑理念是通过反映自身源远流长的文化、根深蒂固的历史以及源于其中的平等权利，来象征土耳其。

土耳其展馆包括了室外、半露天及室内场馆，根据安纳托利亚的不同文化味道进行设计。

展馆入口采用Seljuki星牌向游客问好，其建设利用了大理石、泥土、玻璃、陶器、铜、钢材以及粉饰，并且所有这些材料均为土耳其建筑所独有。

土耳其展馆的屋顶同时代表着过去及未来。阳光穿过最优质的透光材料，创造出明亮宽敞的环境。

土耳其展馆通过摩登风格，重新诠释过去应用在穆斯林庭院和清真寺中传统的建筑设计手法。

地板采用不规则的石头设计，代表着Ephesus（以弗）的路径。这条道路寓意着古老的Anatolian（安纳托利亚）文明之旅。

能容纳上百游客的半露天空间设计，灵感源自穆斯林及清真寺的"庭院文化"，是土耳其展馆兼收并蓄的设计。该区域能够容纳上百名游客。

为了展示土耳其与水的紧密联系，展馆中设有倒映着天空的装饰水池。

展馆内通过大型的通风设备，允许风和阳光进入，形成"自然气候"。凭借此环保特征，土耳其展馆从各国家的展馆中脱颖而出。

建筑理念反映了土耳其文化的全部多样性，突显出风格、知性以及概念丰富程度。

展馆中摆设着以现代手法表现的、过去常用的篮子与泥罐。通过这些展品强调"保护、维持以及再利用"概念。

通过几何图案及色调表达了土耳其文化中深邃而多层次的一面。这些几何图案与形式，根据太阳的运动，在地面上创造了不同的美学阴影图样。

Our architecturalapproachis to symbolize Turkey through reflecting our long-standing culture, deep-rooted history and even-handed power emanating from them.

Consisting of outdoor, semi-outdoor and indoor venues, the Turkish Pavilion was designed based on the flavours of different civilizations that have existed in Anatolia.

Greeting its visitors with Seljuki Star at the entrance, the Turkish Pavilion was built by using marble, earth, glass, ceramic, copper, steel and abode as well as embroidery, all of which are used ingeniously in the Turkish architecture.

The roof of the Turkish Pavilion was designed in a way to represent the past and the future simultaneously. A bright and spacious ambiance lightened through sunlight was created through the best quality light-transmitting materials.

Traditional architectural approach that was used in courts of madrasas and mosques in the past were reinterpreted with a modern style in the Turkish Pavilion.

The floor was designed to represent the Ephesus path made of amorphous stones. The path signifies the journey of the Anatolian civilizations from the antiquity to the present.

Through inspiration from the 'court culture' of our madrasas and mosques, a semi-outdoor space, which can host hundreds of visitors, was designed to contribute to the eclectic design of the Turkish Pavilion. This area can host hundreds of visitors.

In order to showcase our close relationship with water, decorative pools that reflect the sky were installed in our pavilion.

The areas within the Pavilion were equipped with 'natural climatization' through large vents allowing wind and sunlight in. With this environment-friendly feature, the Turkish Pavilion differentiates from other countries' pavilions.

Our cultural diversity was reflected in its entirety in our architectural approach to underline our stylish, intellectual and conceptual richness.

Modern representations of such utensils as baskets and earthen jars that were commonly used in the past were included inside the Pavilion. The concepts "preserve, sustain and recycle" were emphasized through these representations.

The deep and multi-layered aspect of our culture is represented with geometrical motifs and colors. These geometrical motifs and forms create different aesthetic shadowed patterns on the ground according to the movement of the sun.

454
IDEA-TOPS
艾特奖

NOMINEE FOR BEST DESIGN AWARD OF DIGITAL ARCHITECTURE

最佳数字建筑设计奖提名奖

袁烽（中国上海）

获奖项目/Winning Project

西岸FAB-UNION SPACE
FU SPACE on the West Bund

设计说明/ Design Illustration

本项目由袁烽和上海创盟国际建筑设计有限公司（以下简称"创盟国际"）建筑师设计，是城市中令人印象深刻的设计实践。虽然只是小规模工程设计，但是FU Space却充分展示了其建筑价值转移的全新态度。该项目位于上海西岸地区，该地区未来将被规划成上海市艺术文化中心。项目毗邻香格纳画廊、上海展览中心和其他几个建筑师工作室。项目地理位置极佳，离黄浦江畔仅有200m的距离，离龙华寺仅隔两个历史街区。此外，龙美术馆、上海摄影艺术中心和余德耀美术馆也近在咫尺，可短距离步行到达。这个项目无疑将成为一个艺术社区，是具有指导意义的上海艺术尝试。
未来，FU Space将成为非盈利的当代艺术、建筑和文化交流中心，同时有望成为艺术展览和交流平台。
本项目所在地正好处在拐角处，看起来十分紧凑。设计时，充分考虑到了三个不同的流通方向，与上海展览中心二楼平台对接。因此，项目设计的基本概念就是要为整个社区创建一个美好而柔和的接合点。经过认真分析之后，找到了项目所在地的几何形状结构。前述柔和效果主要通过材料来体现，灵感源自平台混凝土结构，混凝土具有色调冷艳和外观粗糙的特点。如果采用模具系统，我们便可将混凝土应用至柔和表面中。
项目专为展览而设计。五个基本空间包括两个4.2m高的空间和三个2.8m高的空间。这些空间呈正四边形结构，可提供灵活和多功能用途。所有特殊的空间体验主要体现在相互之间的流通空间中。室内公共空间与室外结构相呼应。通过抽象思维想像出中国园林假山攀爬的体检，进而完善这种空间结构。中国园林设计的主要特点就是以小衬大。这种做法在景色视点变换中起到十分有效的效果。这种体验的抽象过程通过HP表面几何图形来实现，从不同角度对其进行完善。
施工历时非常短。虽然混凝土墙上出现了不良标记和痕迹，但多样化为紧凑空间带来了全新的空间感。混凝土的柔和可以让光线悠然照射，深深触动参观者的内心深处。
FU Space的初次展览体现了首席建筑师袁烽的设计思维、工作模式以及"缺席的存在"主题。希望这将是对建筑概念的完美诠释。
设计构思首先保证两侧的双层高展厅与三层低展厅相对完整，而中间仅有3m的交通，该空间的构思是建立在人的动态行为、空气动力学拔风以及最大化空间体量连续性基础之上的。
动态非线性的空间生形是建立在结构性能优化以及空间动力学生形的基础之上的，整个过程运用了切石法和投影几何以及算法生形的多种设计方法。
混凝土作为可塑性材料承载了建构特性，同时又具有异域施工的特点，整个建筑从设计到施工历时仅四个月，应该是数字化设计以及施工方法带来的奇迹。

The project which is supposed to be an impressive practice in the city is designed by Philip F. Yuan and Archi-Union Architects. Although it is micro in scale, FU Space is powerful enough to represent a new attitude to the value shift in architecture. It is located in the West Bund area in Shanghai, which is planned to be a future art and culture center of the city. Adjacent to ShangART Gallery, Shanghai Art fair Center and several other architect studios, the location of the site is terrific It is only 200 meter away from Huangpu river front and 2 blocks away from the historical quarter of Longhua Temple. Moreover, Long Museum, Shanghai Photography Museum and Yuz Museum are all within walking distance. It's undoubtedly among the art community which is taking significant experiments in Shanghai.
FU Space will become a future non-profit contemporary art, architecture and culture communication center. It is aimed to be an exhibition and communication space.
The site is very compact, located at a sharp turning corner. N Both the different circulations from 3 directions and a connection to the 2nd floor platform of Shanghai Art Fair Center have to be considered. Therefore, the primary concept is to set up a good soft joint for the whole community. The analysis leads to a form finding process throughout the geometry of the site. The inspiration of the material to reach this softness comes from the concrete of the platform, which originally cold and tough. If we use the mould system, we can actually implement the concrete to any soft surface.
The program is specially set for exhibition. Five basic spaces including two 4.2M height space and three 2.8M space are all regular square ,which could be flexible for multi-functional purposes. . All special space experience lies in the in-between circulation space. Interior public space followed by the exterior form finding process, is enhanced by an abstract thinking on creating a kind of experience climbing the rockery of Chinese garden. The key aspect of Chinese garden design is to make it big through small scale, which is extremely efficient in changing sceneries with varying viewpoints. The abstract process for this kind of experience is achieved by the HP surface geometry, which is intensified from different perspective.
The construction process was conducted in a very short time. Although all the unsatisfied marks and traces were recorded on the concrete wall, a real diversity strengthened a new sense of place in

such a compact space. The softness of concrete, which makes the light flowing leisurely, touches the depth of heart of the visitors.
The first exhibition of FU Space is especially dedicated to the chief architect, Philip F. Yuan exhibits his thinking, working models, and a topic: Presence of Absence, hopefully that could be an explanation for the concept of the building.

The primary consideration in the conception of this project is to ensure theexhibition buildings on the side relatively complete. But the road space of 3meters is built on the basis of the dynamic behavior of people, the air dynamics of the wind and the maximum volume of space continuity.
Dynamic nonlinear spatial shape is built on the basis of structural performanceoptimization and spatial dynamics. The whole process use a variety of design methods of cutting stone , perspective geometry, and the Algorithm configuration.
As plastic material, concrete has the characteristics of construction and has the characteristics of foreign construction, the whole building from design to construction lasted for only four months should be a digital design and construction method of the miracle.

458

IDEA-TOPS
艾特奖

459
艾特奖

主要作品 THE MAIN WORK

2015年，梧桐山中視豐德影視基地（城市復興）
2015年，深圳喜之郎總部大樓（總部大樓）
2014年，深圳瑞和總部大樓（舊樓改造）
2014年，廣州大佛寺佛教藝術大樓（寺廟建設）

HUANG JUN HUI

黄俊琿

深圳瑞和建築裝飾股份有限公司設計院執行院長
深圳市黄俊琿主題酒店設計有限公司董事
深圳皇馬設計師俱樂部創始人
深圳市藝覺設計有限公司董事
深圳市貝殼軟裝網創始人
深圳市江蘇商會會員

ARTHOUSE
Design Group

深圳市罗湖区宝岗路269号大华大厦5楼502
深圳市艺居软装设计有限公司 高端软装顾问机构
Tel:0755-25905829
www.Art-HouseDesign.com
E-mail:ArtHouse2009@126.com

艺居软装 / ARTHOUSE

熙舍

四维吉祥空间设计研习院

重溯传统文化之美
寻找人类灵魂的栖息地

【释】四维吉祥空间设计

溯传统文化之五行学说、方位学说、吉祥文化及东方美学意境之四个维度,以"吉祥环境养生学"之理念营造"天人合一"的厚生空间。

营造范围

大吉祥住宅、别墅空间设计,东方禅意空间室内设计,软装氛围设计,主题文化空间设计、策划,图腾设计⋯

营造观念

以古驭今——传承观
以境载道——空间观
天人合一——宇宙观

熙舍四维吉祥空间设计研习院

吉祥号/400-080-3818

体验馆/深圳市福田区红荔路花卉世界2街75B 熙舍

. 域道设计 .

. 域道设计旗下品牌 .

翁永军
Weng Yong Jun

设计师 | 艺术家 | 收藏家 | 旅行家

广州市筑意空间装饰设计工程有限公司董事长、总设计师；
2001年创立广州集英社设计工程有限公司；
2005年改名广州市筑意空间装饰设计工程有限公司。
现属下公司：
贵州集英社设计工程有限公司；
广东省八建集团装饰工程有限公司贵州分公司；
钧天国际规划设计院(香港)；
广州钧天建筑规划设计有限公司；
广州澳园景观规划设计有限公司；
广州钧天静池古琴文化有限公司。

贵州保利公园2010北区售楼部

Zhu Yi Art

电话：020-34337969
地址：广州市海珠区新港东路1020号保利世贸中心D座45层

电话：0851-85266182
地址：贵阳市乌当区顺海中路88号保利温泉新城商业街1期2楼14-18室

手机版网站：zhuyiart.com
电脑版网站：www.zhuyiart.com
电子邮箱：952600340@qq.com

HRD
HUGE ROCK DESIGN 吴 文 粒 设 计 事 务 所
HTTP://WWW.HRDSZ.COM   WU WENLI DESIGN FIRM

盘石创意·定制级设计

地址：深圳市南山区侨香路与深云路交汇香年广场B座201
T: 0755-2220 0009
E: PS1917@163.COM

设计总监

执行总监

YOUDO DESIGN

广州尤度设计有限公司

A: 广州市海珠区新港东路238号
世港国际G1栋301
T: 020-8955 7899
H: www.youdu999.com
E: youdu88@126.com

朱厚铭

出生于1977年，毕业于广州美术学院环境艺术设计系，广州尤度设计有限公司首席设计师，顶创设计机构董事，中国建筑学会室内设计分会会员，设计专业委员会委员，广州市设计产业学会理事，尤度东方文化传播主持人，**潮菜酒楼专业设计师与策划人。**

尤度设计将一如既往坚持现代建筑语言与人文精神的融合与创新，体现东方特色与西方文化相结合的精巧设计风格，构筑艺术与实用功能相结合的精神空间，致力于研究空间设计的独特性和富有文化内涵的设计作品。

专注 **地域文化餐厅设计&策划**

专注于酒店设计、商业地产设计、公共建筑空间设计、商业空间设计
Http://www.szmok.com

深圳墨客环境艺术设计有限公司
Tel:0755-86096661 / 设计部 E-mail:185792468@qq.com / 地址:深圳市南山区华侨城创意园A4栋303

FSC国际认证　170个国家专利　国家高新技术企业

松博宇科技股份有限公司
国际影响力的新型装饰板材生产基地

「松博宇」、「森驰」、「奥尔伍德」三大产品品牌

3D板丨艺术板丨生态有机板丨美化木四大核心产品系列

1000多个品种，广泛用于专业定制衣柜厨柜、家私饰面、酒店、礼堂等公装及家装领域

———

创新设计/健康环保/艺术个性

实现每一位设计师与消费者对空间的多元化完美期待

深圳市松博宇科技股份有限公司
SHENZHEN SONGBOYU TECHNOLOGY CORPORATION

地址：广东省深圳市宝安区松岗街道沙浦洋涌工业区
电话：0755-29932018
传真：0755-29932008
邮箱：sby@szsby.com
网址：www.szsby.com

before

now

积木易搭
改变，从今天开始

空间大师，3D轻松整家设计；
自带灯光，自带渲染；
硬装软装，数十分钟轻松呈现。

极速720度3D动态云渲染；
视角、动态渲染数十秒同步完成。

配饰助手，2D图片瞬间转3D；
产品无需建模，3D空间软装方案直观极速体验。

一键生成报价清单，在线购买，
轻轻松松一"鼠"搞定，
数百倍地提高工作效率，
数十倍乃至数百倍地提高收益。
为设计师、材料商、家装公司、消费者等
关联产业构筑起无缝对接的平台。

Jmyida.com

深圳积木易搭科技技术有限公司
电话：
0755-23910066
地址：
深圳市福田区福华三路卓越世纪中心3号楼3107

积木易搭 3D云设计产业平台

改变，从今天开始

积木易搭——致力于以技术驱动设计创新，推动全民设计为目的的3D云设计产业平台。

积木易搭采用全球领先的、人性化的互联网3D云设计技术，简单、方便、直观、快捷，数百倍地提高工作效率，彻底解放设计师。以云设计为入口，建立新的体验场景，为设计师、材料商、家装公司、消费者等关联产业构筑起无缝对接的平台；突破时间与空间的局限，改变传统的沟通和销售模式，让信息与需求无缝对接。通过聚焦设计释放整个泛家居产业链的能量，借助互联网的翅膀实现传统家居行业向轻资产、精定位、高增长的转型。

For 设计师

设计师线上独立品牌馆

- 提供强大的设计师个人后台品牌馆管理系统，可以快速上传和管理设计方案、产品素材以及720度3D全景工作室；
- 设计师可在品牌馆中上传展示自己的设计作品，供消费者浏览并在线购买；
- 设计师可在品牌馆上传自己的产品素材，运用积木易搭的3D云设计工具进行方便、快捷的方案设计；
- 设计师可在品牌馆中建立自己的产品资源库并进行产品管理。

线上虚拟720度3D全景工作室

- 设计师可入驻线上虚拟3D"积木大厦"，拥有线上720°全景工作室；
- 设计师可将现实中的办公空间搬到线上，也可设计线上虚拟工作室，让广大有梦想的设计师实现"创客"身份的转型。

深圳积木易搭科技技术有限公司

电话：
0755-23910066

地址：
深圳市福田区福华三路卓越世纪中心3号楼3107

深圳国际家居装饰博览会
家居中国（深圳）创意设计周

3.7-10 / 8.7-9

展览地点：深圳会展中心（全馆开展）

扫一扫！加微信了解更多。

HOME FURNISHING EXPO
Shenzhen China

主办单位：全国工商联纺织服装业商会
　　　　　广东省家用纺织品行业协会
　　　　　中国家用纺织行业协会布艺专业委员会
承办单位：深圳市博奥展览有限公司

地址：广东省佛山市禅城区汾江南路金源街33号世纪嘉园5-202
电话：0757-82363291　82363292　传真：0757-82363290
网址：www.a2398.com　www.hometextiles.cn
请关注官方微博：@广东家纺协会　@家居中国-深圳家纺展

ONLEAD
人性空间 自由办公

FREE WORK & HUMANISTIC SPACE

广州市欧林家具有限公司

欧林是中国首批办公家具制造商之一，
1996年成立以来，一直专注于办公家具研发、生产及销售，
致力于为客户提供专业的办公空间整体解决方案。
欧林以创造人性化的办公空间为使命，
成功为众多世界500强企业提供了优质服务，
销售网络覆盖全国50多个大中城市，
是中国办公家具行业领导品牌。

CONTACT US

400 8868212

✆ +86-20-6660 1678
✆ +86-20-6683 0222
✉ www.onlead.com.cn
📍 广东省广州市白云区人和镇鹤龙7路100号欧林工业园

Website　　Wechat

高档装修
不用大理石，就用简一®

简一® 大理石瓷砖
出口意大利、法国等60多个国家

年鉴
WORKS OF
OPS

CONTENTS 目录

006/ 中国境内国际化程度最高的
专业设计大奖Idea-Tops艾特奖
Idea-Tops, the Most Internationalized
Professional Award for
Design in Mainland China

008/ 序言
Preface

014/ 2015 Idea-Tops艾特奖评审委员会
Idea-Tops Jury Panel 2015

016/ 设计师们眼中的艾特奖
Idea-Tops in the Eyes of Design Masters

022/ 2015年度艾特奖颁奖盛典
Idea-Tops Awarding Ceremony 2015

034/ 第二届G10设计师峰会
The 2nd G10 Designers' Summit

040/ 艾特奖国际学术委员会
Idea-Tops International Design Forum

044/ 2015艾特奖国际大师论坛 拉近
中国与世界的距离
The 2015 Idea-Tops International Master
Forum Shortened the Distance between
China and the World

050/ **A** 最佳别墅设计奖
Best Design Award of Villa

074/ **B** 最佳酒店设计奖
Best Design Award of Hotels

094/ **C** 最佳会所设计奖
Best Design Award of Clubs

116/ **D** 最佳样板房设计奖
Best Design Award of Show Flat

140/ **E** 最佳公寓设计奖
Best Design Award of Apartment

164/ **F** 最佳办公空间设计奖
Best Design Award of Office Space

188/ **G** 最佳娱乐空间设计奖
Best Design Award of Entertainment Space

212/ **H** 最佳餐饮空间设计奖
Best Design Award of Dining Space

236/ **I** 最佳展示空间设计奖
Best Design Award of Exhibition Space

264/ **J** 最佳文化空间设计奖
Best Design Award of Cultural Space

284/ **K** 最佳商业空间设计奖
Best Design Award of Commercial Space

308/ **L** 最佳光环境艺术设计奖
Best Design Award of Lighting Design

332/ **M** 最佳陈设艺术设计奖
Best Design Award of Art Display

360/ **N** 最佳住宅建筑设计奖
Best Design Award of Residential Architecture

384/ **O** 最佳公共建筑设计奖
Best Design Award of Public Architecture

412/ **P** 最佳绿色建筑设计奖
Best Design Award of Green Architecture

436/ **Q** 最佳数字建筑设计奖
Best Design Award of Digital Architecture

Idea-Tops, the Most Internationalized Professional Award for Design in Mainland China

中国境内国际化程度最高的专业设计大奖
IDEA-TOPS艾特奖

国际空间设计大奖——Idea-Tops艾特奖,是中国境内国际化程度最高的专业设计大奖,建基于全球第二大经济体及迅猛发展的设计市场,旨在发掘和表彰在技术应用、艺术表现及文化特质再现上,具有创新意识的设计师和设计作品,打造全球最具思想性和影响力的设计大奖。

艾特奖为推动中西方设计交流搭建了一个沟通协作的平台,高水平的国际级评委、严谨公正的评奖机制及奖项设置……使艾特奖成为了中国境内最具国际化和专业性的设计奖项,也成为了世界设计业了解中国设计的一个窗口。

艾特奖参与者可谓众星云集,包括全球三大设计事务所之一的Gensler设计总监Graeme Scannell、"中国第一高塔"广州塔设计者Mark Hemel,全球酒店设计公司五强、BBG-BBGM建筑与室内设计公司设计董事Robert J.Gdowski,拥有140年历史的国际知名建筑事务所Woods bagot全球总监Rodger Dalling;英国首相官邸——唐宁街10号设计者、BBC苏格兰总部设计师Ross Hunter,希尔顿国际酒店集团主创设计师Martin Hawthornthwaite;曼联俱乐部亚太区首席设计师Mike Atkin;深圳机场T3航站楼设计者FUKSAS夫妇;2008年普利兹克奖获奖者、法国当代著名建筑师让·努维尔(Jean Nouvel)等。此外,2015年米兰世博会中国、意大利、土耳其、奥地利、阿尔及利亚、捷克、摩尔多瓦等11大国家馆强势竞逐2015艾特奖,再掀业界风云。

源于东方,面向世界,Idea-Tops艾特奖崇尚的,是设计师永不枯竭的智慧与前瞻性的革新思想,以及他们审美和工艺上均显卓越的设计作品。恪守专业、严谨、公平、公正的原则,Idea-Tops艾特奖以至高的专业标准、专业发展、专业责任以及专业沟通来促进设计业的发展,每届艾特奖均邀请全球设计领域资深专家、学者、顶尖建筑师和设计师、知名人士、财经专家、意见领袖以及有影响力的媒体担纲评委。

作为表彰建筑和室内设计界杰出人才的重要奖项,获得艾特奖也就是向全球展示了精英们在建筑和室内设计界的顶级荣誉。在中国,超过2/3的知名房地产开发将艾特奖获得者锁定为首选合作伙伴。在欧美、亚太地区,艾特奖正逐步成为设计师

International Space Design Award—Idea-Tops is the most internationalized and professional design award in China, its establishment was based on the rapidly developing design market of the world's second-largest economy. Idea-Tops aims to discover and praise designers and design works that are innovative in the space form, technology application, art performance and culture presentation, and create the most thoughtful and influential design award around the world.

Idea-Tops builds a platform for communication and collaboration between Chinese and Western designs. The high level of international judges, rigorous and impartial mechanism of awards and the setting of award categories make Idea-Tops not only become the most internationalized and professional design award in China, but also a window for the world design industry to understand Chinese design.

Lots of masters participated in Idea-Tops, including Graeme Scannell (design director of Gensler—one of the world's top three design studios), Mark Hemel (designer of Guangzhou Tower—China's tallest Tower), Robert J.Gdowski (designer principal of BBG-BBGM—one of the world's top five hotel design companies), Rodger Dalling (global director of Woods Bagot—the international renowned architecture firm which has a history of 140 years), Ross Hunter (designer of No.10 Downing Street and BBC Scotland headquarters), Martin Hawthornthwaite (chief designer of Hilton Hotels Corporation), Mike Atkin (Asia-Pacific principal designer of Manchester United Football Club), Massimiliano and Doriana Fuksas (designers of Shenzhen T3 Airport), Jean Nouvel (winner of 2008 Pritzker Architecture Prize, famous contemporary architect in French), etc. In addition, 11 national pavilions from Expo Milano 2015 competed in Idea-Tops 2015 forcefully, such as China pavilion, Italy pavilion, Turkey pavilion, Austria pavilion, Algeria pavilion, Czech pavilion and Moldova pavilion, etc. which evoked strong repercussions in design industry.

Idea-Tops originated from the East and faces to the world, what it underscores are designers' inexhaustible wisdom and forward-thinking innovation as well as their works which are excellent in aesthetics and craft. With the principles of being professional, rigorous, impartial and fair, Idea-Tops promotes the development of design industry by professional standards, professional development, professional responsibility and professional communication. In each session, Idea-Tops invites a fantastic and distinctive judge panel, including senior experts, scholars, top architects and designers, celebrities as well as financial experts, opinion leaders and influential media representatives in the global design field.

Idea-Tops is an important award which praises outstanding designs, getting its recognition is a global announcement of designers' great honor in the architectural and interior design field. In China, over two-thirds of renowned real estate developers will choose Idea-Tops awardees as preferred partners. Idea-Tops is becoming a hot topic in the design community in

007

IDEA-TOPS
艾特奖

未来，已来

上帝设计了世界
小至沙尘，大至江海
他们设计了生活
小至寓所，大至国城
他们
不仅定义了我们的空间
更定义了我们的未来

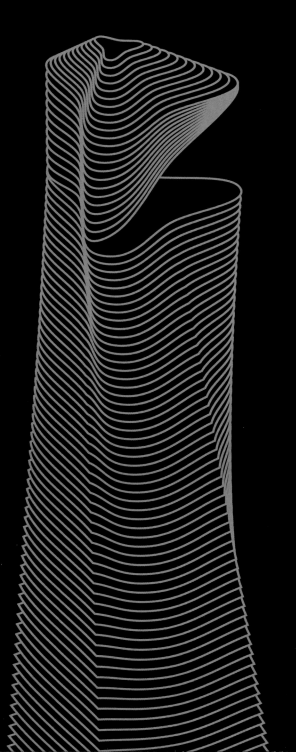

21世纪的设计探索
Design Exploration in the 21st Century

刘育东

艾特奖国际学术委员会副主席兼执行主席
台湾亚洲大学副校长、哈佛大学建筑学博士

Liu Yu-Tung, Vice Chairman and Executive Chairman of Idea-Tops International Design Forum, Vice President of Asia University, Taiwan and Doctor of Design in Harvard University

21世纪是一个非常大的时代,我们常常思考中国在世界最重要的设计和建筑领域需要探讨的当代议题。20年前,我们想象这个名词非常前卫,好像是电影或动画片中描绘的天马行空式的前卫。但是2015年告诉我们,事实上我们已经在21世纪的前期,而且已经经过了15年,我们可以看到许多建筑大师在21世纪初期,建立了一个全新的建筑与空间、建筑与城市、建筑与环境的关系,但是我们对21世纪之设计的探索却仍然不够,尤其是数字化科技对未来的建筑设计产生的影响。

受赵会长的邀请,我从3年前开始参加艾特奖,3年来看到大量反映当代设计思维的作品。一方面,这些作品融合了具有深厚传统内涵的元素,把这些元素变成在空间上、设计上、甚至在外观上重要的视觉展示内容。另一方面,它们又很全球化,因为这里面有很多来自世界各地,很新颖、很现代的概念与素材,呈现出一种中西相融的非常蓬勃的设计发展趋势。大家都在设计,而设计是一件看未来的事情,因此我们非常有必要探讨在21世纪余下的85年里应该走哪个新的方向。

自去年开始,我和赵会长即策划以艾特奖为平台,吸引中国两岸四地乃至全世界重要学府的学者教授一起探讨:基于当代建筑与社会的发展,我们需要一个什么样的属于21世纪的设计理念,这促成了艾特奖国际学术委员会的诞生。在此,特别感谢上海同济大学郑时龄院士出任艾特奖国际学术委员会主席,拓宽我们的视野并全面提升我们的学术高度。感谢清华大学美术学院鲁晓波院长共任学术委员会副主席,并参与支持委员会各项工作。亦感谢赵会长的团队,让我们汇集哈佛、牛津、剑桥、耶鲁、麻省理工、普林斯敦等世界各国顶尖学府的专业学者,同台分享与交流位于时代最前沿的设计理念。

2015是起点,在未来的发展中我们还要举办85届艾特奖,到了2099年,21世纪的最后一天,我相信艾特奖国际学术委员会将被载入21世纪史册。同时,每年的艾特奖论坛与学术期刊,将持续聚焦未来每个年度有关21世纪大家应该正视的问题与应该解决的问题,为这个世纪设计的发展提供更多的可能性。

The 21st century is a great era. We often consider what contemporary issues China needs to be discussed in the most important design and architecture field of the world. Twenty years ago, the 21st century was very avant-garde to us, just like the unrestrainedly style in movies or animations. But in fact, we're living in the 21st century now, and we have seen many architectural masters establish a new relationship between architecture and space, architecture and city, architecture and environment in the past 15 years of the century. However, our exploration on design is far from enough, and the impact of digitalized technology on architectural design deserves particular attention.

Being invited by Chairman Zhao, I began to participate in Idea-Tops three years ago. For these three years, I have seen a lot of works reflecting the contemporary design thinking. On the one hand, they have been integrated with elements that have profound traditional connotation. These elements were transformed into important visual display contents in space, design and even appearance. On the other hand, they were rather globalized since there were many new and modern concepts and materials from all over the world, presenting a very prosperous design development trend of combination of Chinese and western elements. Everyone is designing, but design is future-oriented. Therefore, it is very necessary for us to discuss what direction we should take in the rest of the 21st century.

From last year, Chairman Zhao and I began to work on such a plan that takes Idea-Tops as a platform to attract scholars and professors from important universities in China and the world to discuss a question—based on the contemporary architectural and social development, what kind of design concept do we need in the 21st century? This contributed to the establishment of Idea-Tops International Design Forum. On this occasion, I express my special thanks to Zheng Shiling (Academician of the Chinese Academy of Sciences, Professor of Tongji University) who serves as chairman of Idea-Tops International Design Forum, expands our vision and fully improves our academic level. I thank Lu Xiaobo (President of Academy of Arts & Design, Tsinghua University) for assuming the role of vice chairman of Idea-Tops International Design Forum and participating in all related tasks. I also thank Chairman Zhao's team for gathering professional scholars from the world's top universities such as Harvard, Oxford, Cambridge, Yale, MIT and Princeton to share and exchange today's cutting-edge design concepts.

The year of 2015 is a starting point. We will hold 85 sessions of Idea-Tops in the rest of the century. On the last day of 2099, I believe that Idea-Tops International Design Forum will be written down in the history of the 21st century. Meanwhile, we will always focus on the problems that we should face and solve every year through this forum and academic journals, and provide more possibilities for the development of design in this century.

这是个挑战，也是个机会
This Is a Challenge and an Opportunity

鲁晓波 / 2015艾特奖国际评审委员会主席
清华大学美术学院院长、博士生导师
Lu Xiaobo, Chairman of International Jury in Idea-Tops 2015, President and Doctoral Supervisor of Academy of Arts & Design, Tsinghua University

很荣幸受邀参与2015国际空间设计大奖——艾特奖的终评工作，与来自牛津大学、罗马大学、巴黎高等建筑学院、麻省理工大学与台湾亚洲大学的专家教授共同评选本年度最优秀的建筑室内设计作品。

本次参赛的作品来自世界各地，涵盖很多领域。大家在环境保护、可持续发展、特定功能、特定空间的研究，以及艺术形式、传统文化、更健康生活方式塑造方面都体现了深思熟虑的设计逻辑，涌现了不少相对来讲具有创新意义的前沿设计概念，或者是设计解决方案，给我留下了非常深刻的印象。

尤其让我感到惊讶的是，这次有十余件2015米兰世博会国家馆的建筑方案参赛。这一方面是个信号：虽然我们的房地产在降温，国外一流的设计公司仍然看好中国未来的设计市场；另一方面，也代表艾特奖的影响力在提升，大家正逐渐认同这样一个重要的奖项，我想这对于中国提倡创新型国家建设，强调我们的创新能力建设，具有非常重大的意义。

同时，我也看到这么多奖项里面，各个类别之间的差异性不够，说明我们还可以在功能研究上多下工夫。我期待下一届会有更多更具特色、更具原创性的作品出现。我们的设计师可以借助这样一个调整时期充实自己，更多了解前沿科技发展对这个行业产生的影响，充分挖掘自己民族深厚的文化底蕴，真正设计出引领性的创新型作品，或者在自己有限的设计项目中潜心研究、精心设计，打造一批更好的传世之作。

我认为，这是个挑战，也是个机会。

I am honored to be invited to participate in the review of International Space Design Award—Idea-Tops in 2015 and evaluate the most excellent architecture and interior design works of the year with experts and professors from University of Oxford, University of Rome, The Superior National School of Architecture Paris-Val-de-Seine, MIT and Asia University, Taiwan.

The entries were from different places of the world and covered many categories. Deliberate design logic was reflected in environmental protection, sustainable development, particular functions, specific space research, artistic form, traditional culture and healthy lifestyle. It really impressed me that there were a lot of innovative and cutting-edge design concepts or design solutions.

In particular, I'm very surprised that over ten architectural projects of national pavilions from 2015 Milan World Expo competed in Idea-Tops. This is a signal: Although our real estate market is cooling, some first-class foreign design companies are still optimistic about China's future design market. It also manifests that the influence of Idea-Tops is increasing and it's being recognized as an important award. I think that this will be of great significance for China to advocate innovative country construction and stress on our innovative capacity.

At the same time, I realized that there is not much difference among the awards, and this indicates that we can do more on the research of function. I hope more special and original works will appear in the next competition. Designers can make use of this period to adjust and enrich themselves, learn more about the impact of cutting-edge technical development on the industry, explore profound cultural heritage of their own nations, and design leading and innovative works or concentrate on researching and designing their limited projects so as to create better works.

I think that this is a challenge and an opportunity.

011
IDEA-TOPS
艾特奖

让中国设计成为在全球受人尊敬的设计力量
Make Chinese Design a Design Force Respected Worldwide

赵庆祥 / 艾特奖组委会执行主席 、 深圳市政协委员
Zhao Qingxiang, Executive Chairman of Idea-Tops Organizing Committee and Committee Member of CPPCC in Shenzhen

作为中国设计行业的标杆奖项，2015艾特奖全球巡回推广活动自2015年3月初就拉开了序幕，历时8个月，走过国内外25个城市，举办了32场大型的设计论坛，48场高规格的设计沙龙，近5万名设计师参与，跨越国界、跨越大江南北、跨越海峡两岸，正是这样一场史无前例的巡回推广活动，为不同地区的设计师、设计机构搭建了一个国际化的学习平台，助推了中国设计业的发展，让一批真正有实力的设计公司用作品说话，迅速成长，让更多有思想、有潜质、有梦想的设计师们看到了希望。

今年，艾特奖组委会共收到来自英国、美国、意大利、德国、法国、希腊、印度、日本、新加坡以及中国等35个国家和地区的参赛设计作品，共5682件，无论参赛作品的数量，还是作品质量，均创历届之最。所以，今年的获奖者实属不易，所有获奖作品经过入围奖评选、提名奖评审以及国际终评。3个环节的严格评审，在艾特奖这个专业、严谨、公正的平台上，从众多的作品中脱颖而出，本届获奖的设计师们是当之无愧的最值得尊敬的设计明星。

特别值得一提的是，艾特奖组委会携手22所世界顶级大学(包括哈佛、耶鲁、剑桥、普林斯顿、麻省理工学院、清华、北大等)的知名教授及学术领军人物组织并成立了艾特奖国际学术委员会，由中国科学院院士郑时龄担任学术委员会主席，由清华大学美术学院院长鲁晓波担任副主席，哈佛大学建筑设计博士、台湾亚洲大学副校长刘育东担任副主席兼执行主席，并召开了首届国际学术会议，这是艾特奖登顶设计领域国际学术新高度、助力设计产业创新的一座里程碑。它的成立不仅可以进一步完善艾特奖的专业标准、学术理论体系，更可以用学术力量掀开中国设计新的一页，为中国乃至世界设计业的进步，新技术、新材料、新零件的运用，提供源源不断的理论探索、智力支持和学术成果支撑。

站在新的起点上，艾特奖将以更高的专业标准、专业发展和专业沟通，让中国设计成为在全球受人尊敬的设计力量。

As a benchmark in the Chinese design industry, Idea-Tops started its global promotion tours at the beginning of March in 2015. During the 8 months, Idea-Tops has held 32 large-scale design forums and 48 high-standard design salons in 25 cities at home and abroad. Nearly 50,000 designers participated in these activities. Being spread all over the world, the unprecedented promotion campaigns built up an internationalized learning platform for designers and design institutions from different regions. This has promoted the development of Chinese design industry, made a batch of powerful design firms grow rapidly with their works, and gave hope to designers who have thoughts, potentiality and dreams.

This year, Idea-Tops Organizing Committee has received 5,682 works from 35 countries and regions, such as UK, USA, Italy, Germany, France, Greece, India, Japan, Singapore, Chinese Mainland, Hong Kong, Macau and Taiwan, etc. The quantity and quality of the entries are the highest in history. Therefore, it is not easy for participants to win the awards. Each entry is strictly evaluated through shortlist review, nomination review and international final review. All winners are the most respectable design stars because they can stand out from numerous works at the Idea-Tops platform which is professional, rigorous and fair.

In particular, Idea-Tops Organizing Committee has cooperated with famous professors and academic leaders from 22 of the world's top universities, including Harvard, Yale, Princeton, MIT, Tsinghua and Peking, to establish the Idea-Tops International Design Forum. Zheng Shiling (Academician of Chinese Academy of Sciences) serves as chairman of the Idea-Tops International Design Forum, Lu Xiaobo (President of Academy of Arts & Design, Tsinghua University) serves as vice chairman, and Liu Yu-Tung (Doctor of Design in Harvard University and Vice President of Asia University, Taiwan) serves as vice chairman and executive chairman. The holding of the first design forum was a milestone which made Idea-Tops reach the new international academic height in design field and facilitated the innovation of design industry. The establishment of the Idea-Tops International Design Forum can not only further perfect Idea-Tops' professional standard and academic theory system, but open up a new chapter for Chinese design with academic strength as well as provide continuous theoretical exploration, intellectual support and academic achievements for design industry of China and the world and applications of new technologies, materials and components.

Standing at the new starting point, Idea-Tops will make Chinese design a design force respected worldwide with higher professional standards, professional development and communication.

2015 Idea-Tops艾特奖 评审委员会

IDEA-TOPS JURY PANEL 2015

1 **鲁晓波**
清华大学美术学院院长

2 **刘育东**
哈佛大学建筑学博士，台湾亚洲大学副校长

3 **ANTONINO SAGGIO**
罗马大学建筑学教授

4 **DAVID HOWARD**
牛津大学城市可持续发展学系系主任

5 **TAKEHIKO NAGAKURA**
麻省理工学院教授

6 **MARTINE BOUCHIER**
巴黎高等建筑学院教授

Idea-Tops in the Eyes of Design Masters

设计师们眼中的艾特奖

2015年12月2日—3日,2015年度艾特奖系列活动在深圳中海凯骊酒店隆重举行。在室内设计行业迅猛发展、机遇与挑战并存之际,2015艾特奖吸引了全球设计的眼光,共同探讨设计,共话设计未来,形成了强大的磁场力和向心力。一个设计大奖为何能形成如此巨大的影响,掀起行业飓风,艾特奖本身有哪些魅力?在国内外设计大师及专家学者眼中,艾特奖究竟是什么样的?让我们走近国际顶尖设计大师,感受艾特奖的真实魅力。

From December 2nd to 3rd, 2015, the series of activities of Idea-Tops were held at Coli Hotel, Shenzhen. At the moment when opportunities and challenges were coexisting in the booming interior design industry, Idea-Tops 2015 had attracted attention from the global design community. People talked about design and future trends, which brought dynamic influence. As a grand award for design, what makes Idea-Tops so influential and appealing? What is Idea-Tops like in domestic and foreign design masters' eyes? Let's find out the charms of Idea-Tops from them.

江湖论道

江湖汇英雄·围坐而论道

南海之滨——"设计之都"深圳，汇集着来自全国乃至全球的优秀设计师。为了给各路英雄提供一个切磋技艺、对话交流的平台，【江西设计·深圳】联合深圳市陈设艺术协会、《设计之都》杂志共同打造"江湖论道"大舞台。其宗旨是：以设计师的角度，深度关注社会热点事件，针对行业内外的热门新闻事件进行探讨，犀利碰撞，各抒己见。每期活动将邀请几位来自设计、文化、家居行业界的知名嘉宾共同论道（嘉宾不分国籍、不分种族、不分男女均可参加），致力于用普适的、主流的、接地气的价值观去进行思想启蒙和价值引导，达到融汇信息传播，制造乐趣与辩析事理三大元素于一身的目的，重新定义设计师。

江西设计·深圳

由位于深圳的江西籍知名设计师团体领衔并筹建的非营利性联合性社会团体
秉承着"**奉献·平等·分享**"的团体宗旨
致力于把 江西设计·深圳 打造成一个有利于设计行业发展
促进设计行业产业链合作并提升设计水平的综合性平台

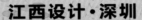

我们同行
奉献·平等·分享
www.jxd0755.com

联系人：王启迪
联系电话：13662265892
地址：深圳市南山区侨香路香年广场A座2001室

GABRIEL LIGHTING

GABRIEL LIGHTING
圣嘉佰利

高端欧美灯饰品牌

灯饰·LED·家具·灯扇·饰品一站式整体服务商

完美石材典范
打造星级石材王国

PERFECT STONE MODEL
BUILD STAR STONE KINGDOM

HC·STONE

　　皇朝石材集团创建于2011年，历经五年多的快速稳步发展，现已发展为一家集矿山开采、多品类石材生产加工、设计开发、安装维护于一体的石材企业。

　　皇朝石材集团拥有一批优秀石材行业精英，可提供工程设计、石材采购、加工、安装维护一站式服务，凭借先进的加工设备、精湛的技术工艺和完善的售后服务赢得客户的好评，迄今已承接深圳T3航站楼、罗湖边检出入境大厦、卓越时代广场、华南城总部大楼、深圳远鹏装饰大楼、金积嘉集团万国食品城、安徽徽商集团酒楼、佳兆业可域酒店、万达西双版纳文华酒店、万达成都瑞华酒店、万达苏州嘉华酒店、武汉白玫瑰酒店、珠海棕泉酒店、广西北海富丽华大酒店、万科双月湾、招商地产伍兹公寓、深圳南岭一半山、新世界地产世纪御园、湖北宜昌恒信中央公园、高邮陆宇中央郡凤凰城别墅、沈阳紫薇仙庄等高级建筑楼宇室内外石材装饰工程项目。

　　展望未来，皇朝人将以"诚信、品质、高效、创新、务实"的精神，全面实施"人才战略"、"品牌战略"、"服务战略"，积极研发独特的技术工艺，探索新的石材营销模式，致力成为全球石材行业发展新风尚的引领者。

香港皇朝石材集团控股有限公司
广东皇朝石材工艺有限公司
深圳市皇朝石材有限公司
深圳市龙岗区南湾布澜大道盛宝路皇朝工业园
TEL：0755-89963888　　FAX：0755-89512555
尊享热线：13026626666　13823237666
www.huangchaostone.com

INTERNATIONAL SPACE DESIGN AWARD

国际空间设计大奖 IDEA-TOPS 艾特奖

8月31日截止 / 3月8日启动

作品全球征集

了解更多资讯请访问艾特奖官方网站：www.idea-tops.com

内容提要

本书为国际空间设计大奖——Idea-Tops艾特奖2015年获奖作品集，内容主要包括最佳别墅设计奖、最佳酒店设计奖、最佳会所设计奖、最佳样品房设计奖、最佳公寓设计奖、最佳办公室空间设计奖、最佳娱乐空间设计奖、最佳餐饮空间设计奖、最佳展示空间设计奖等18项大奖获奖作品的展示与解析，同时对Idea-Tops艾特奖以及2015第二届G10设计师峰会予以简单介绍。

图书在版编目（CIP）数据

2015艾特奖获奖作品年鉴 / 国际空间设计大奖艾特奖组委会编著. -- 北京：中国水利水电出版社，2016.6
ISBN 978-7-5170-4531-1

Ⅰ. ①2… Ⅱ. ①国… Ⅲ. ①建筑设计—世界—2015—年鉴 Ⅳ. ①TU206-54

中国版本图书馆CIP数据核字(2016)第157900号

书　名	**2015艾特奖获奖作品年鉴** 2015 AI TE JIANG HUO JIANG ZUO PIN NIAN JIAN
作　者	国际空间设计大奖艾特奖组委会 编著
出版发行	中国水利水电出版社 （北京市海淀区玉渊潭南路1号D座 100038） 网址：www.waterpub.com.cn E-mail：sales@waterpub.com.cn 电话：（010）68367658（营销中心）
经　售	北京科水图书销售中心（零售） 电话：（010）88383994 全国各地新华书店和相关出版物销售网点
排　版	深圳市东方辉煌文化传播有限公司
印　刷	北京科信印刷有限公司
规　格	210mm×250mm 16开本 30印张 947千字
版　次	2016年6月第1版 2016年6月第1次印刷
印　数	0001—1000册
定　价	598.00元（附光盘1张）

凡购买我社图书，如有缺页、倒页、脱页的，本社营销中心负责调换
（版权所有·侵权必究）